Averting the Digital Dark Age

AVERTING THE DIGITAL DARK AGE

How Archivists, Librarians, and Technologists Built the Web a Memory

IAN MILLIGAN

Johns Hopkins University Press
Baltimore

Printed in the United States of America on acid-free paper

2 4 6 8 9 7 5 3 1

Johns Hopkins University Press
2715 North Charles Street
Baltimore, Maryland 21218
www.press.jhu.edu

Library of Congress Cataloging-in-Publication Data

Names: Milligan, Ian, 1983– author.
Title: Averting the digital dark age : how archivists, librarians, and technologists built the web a memory / Ian Milligan.
Description: Baltimore : Johns Hopkins University Press, 2024. | Includes bibliographical references and index.
Identifiers: LCCN 2024010568 | ISBN 9781421450131 (hardcover) | ISBN 9781421450148 (ebook) | ISBN 9781421450537 (ebook open access)
Subjects: LCSH: Web archiving—History. | Digital libraries—History.
Classification: LCC Z701.3.W43 M55 2024 | DDC 025.042—dc23/eng/20240705
LC record available at https://lccn.loc.gov/2024010568

A catalog record for this book is available from the British Library.

This book is freely available in an open access electronic edition thanks to the support of the Social Sciences and Humanities Research Council of Canada and a subsidy from the University of Waterloo

Special discounts are available for bulk purchases of this book. For more information, please contact Special Sales at specialsales@jh.edu.

For Isla

CONTENTS

ACKNOWLEDGMENTS

One of those annoying—yet true!—administrative clichés is "it takes a village." My sincerest thanks to all who made *Averting the Digital Dark Age* possible. Since 2015 I have been working in the related area of improving access to web archives and thus extend many thanks to my amazing colleagues and collaborators with the Archives Unleashed project that helped me think about web archives every single day: Samantha Fritz, Nick Ruest, Jefferson Bailey, and Jimmy Lin. Thanks also for your patience while I occasionally buried my head away in my mysterious solo project that ultimately became *Averting the Digital Dark Age*.

Big thanks are due to the individuals who agreed to be interviewed for this project and then engaged with me through follow-up emails and conversations. Their names appear throughout the book, but thanks to Edward Higgs, Terry Kuny, Brewster Kahle, Paul Koerbin, Abbie Grotke, Margaret Hedstrom, Steve Schneider, Kirsten Foot, and Tom Scheinfeldt. As I joked to some of them, it is always "fun" writing about events in the recent past. We know, deep down, that historians get all sorts of things wrong, but contemporary historians live with the fear of people who made the histories we write about pointing that out to us. My sincerest thanks to them, and of course, any errors of interpretation are solely mine.

I benefit from having an amazing group of colleagues and friends at the University of Waterloo, both in the Department of History, as well as the Office of the Vice-President, Research and International, where I am currently based. Thanks Charmaine Dean, for encouraging us academic administrators to keep active research portfolios while supporting our researchers across campus! In the Department of History, thanks to our chair, Dan Gorman, for fostering research excellence, as well as many colleagues for fostering such a vibrant, intellectual culture. My thanks also to research assistants Eric Vero, Vera Zoricic, and Rebecca MacAlpine for assisting with either this book or closely related projects.

This book draws on research supported by an Insight Grant from the Social Sciences and Humanities Research Council. We are so fortunate in Canada to be generously supported by these programs.

At Johns Hopkins University Press, my thanks to Matthew McAdam, who not only acquired the manuscript but saw it through the peer review process. This is a difficult undertaking at the best of times, and COVID made editors' lives difficult in so many ways. Thank you sincerely, Matt, for working with me to bringing the manuscript to fruition. My thanks as well to the peer reviewers who provided comments, which greatly improved this book. Special thanks as well for excellent production support by Robert Brown, copyedits by Michael Baker, and indexing by Patti Phillips.

Thanks also to receptive audiences who heard versions of the manuscript presented at invited talks and conference presentations, including the University College London Digital Humanities 2020–21 seminar series, RESAW 2023, and the Between Postwar and Present Day Canada conference. As well, my thanks to Laura Portwood-Stacer's "Book Proposal Accelerator," which gave me the necessary motivation to put the proposal together for this book you are about to read! Special thanks as well to my friend Thomas Peace at Huron University College, who read several of the chapters in this book and always brings an expert eye to many of the things that I write.

Finally, a most important thanks to my family. Much of the manuscript was written during the COVID-19 pandemic, making for an "interesting" time. Isla, Auden, and Jenn: thank you for helping to support the writing of this book. Our family is a great team! You all probably know more about web archiving than you wish you did. Thanks in advance for pretending to be excited when I show you this book on the shelf, Auden and Isla, even though you probably wish it was actually about *Star Wars* or *Frozen*.

Averting the Digital Dark Age

Introduction

In early 1996, the web was ephemeral. By 2001, the web remembered. Today, most of us know that when we share something on the web—a tweet, a blog post, an article, a photo—it can last forever. Yet longevity is not an innate feature of the web. The development of the web's memory was not accidental, nor was it the product of a coordinated master plan. What began as worries amongst research librarians, technologists, futurists, and writers from the Second World War onwards laid the groundwork for today's digital memory. If, in the mid-1990s, commentators worried about a "digital dark age," we are now in an age of historical abundance. Memory institutions preserve petabytes of information every year, much of it generated by ordinary people as they go about their everyday lives.

The specter of a "digital dark age" haunted libraries by the mid-1990s, portending a dark future with no memory. "The digital medium is replacing paper in a dramatic record-keeping revolution," warned Jeff Rothenberg in a January 1995 *Scientific American* article, adding that "such documents may be lost unless we act now."[1] Fears only escalated over the coming years. The combination of more data than ever before being stored, with the need to preserve it over a longer time frame than usual, was, as information scholar Margaret Hedstrom put it apocalyptically in 1997, "a time bomb."[2] Terry Kuny wrote in 1997 at the International Federation of Library Associations conference that to be digital meant "being ephemeral." Kuny predicted that "it will likely fall to librarians and archivists, the monastic orders of the future, to ensure that something of the heady days of our 'digital revolution' remains for future generations."[3] Stewart Brand, former publisher of the *Whole Earth Catalog* and later cofounder of the Long Now Foundation, noted in 1998 that while "we can read the technical correspondence from Galileo . . . we have no way of finding the technical correspondence" of the digital age.[4] To these thinkers and leaders, the digital turn portended our cultural record's destruction.

These commentators articulated fears about the ephemerality of digital information, popularizing conversations that had been percolating around the library and information community for decades. Information has always been fragile, but by the 1960s the prospect of electronic or machine-readable records had increasingly complicated the preservation landscape. Fears of loss accelerated in 1994 with the launch of the joint Commission on Preservation and Access and the Research Libraries Group's Task Force on Archiving of Digital Information, raising the profile of the newly coined "digital preservation" field among research libraries and archives. Networked communication, especially with the rapidly growing World Wide Web, further raised the stakes. The web was part of a much broader context of pressures stemming from changes in digital storage formats, which in a matter of years had seen a transition from large floppy disks to smaller diskettes to CD-ROMs. As personal computing became more accessible by the 1990s, users could see information that had been accessible only a few years ago become lost in a sea of obsolete technology or file formats. The growing number of computer and then web users made the prospect of a digital dark age less science fiction and more a reality.

By the late 1990s, the web's ephemerality became a major challenge as thousands raced to join the "information superhighway." Where there had been under 2,500 websites in 1994, by 1996 there were over a quarter of a million and by 1997 over a million, with the number exponentially growing.[5] Experts and users alike came to see that important websites could suddenly disappear: a server taken offline, a student graduating from college and losing their account, fees going unpaid, and the site—poof!—would be gone forever. The ephemerality of web content was clear. By the mid-1990s, this was affecting more than just a few disappointed geeks. Commentators and technologists saw this as the potential collective imperilment of the human record. A digital dark age would be a calamity. Would the web be ephemeral? Out of need came innovation, represented in part by the 1996 founding of the Internet Archive.

Flash forward a half-decade to the morning of 11 September 2001 in the United States. Hours after hijacked airplanes crashed into the World Trade Center, the Pentagon, and rural Pennsylvania, memory infrastructure sprang into action. The Internet Archive, Library of Congress, and memory institutions around the world acted to preserve thousands of websites, tens of thousands of emails, and other digital artifacts relating to the attacks and their impact. Existing digital infrastructure at these diverse institutions operated to ensure that by the end of the day itself, historians would have hundreds of website snapshots to draw upon to understand that tumultuous day and its aftermath. Only a month after the attacks, on 11 October, a web portal was launched to provide immediate access to a replay interface that let users go "back in time" to websites collected on the day of the attacks. Scholars soon thereafter launched a crowdsourced platform to gather digital information, including digital voicemail recordings, email listservs, and digital photographs. While the record is not complete—commentators bemoaned the loss of rich digital exhibits lost to the obsolescence of Adobe Flash during the twentieth anniversary coverage—it is as comprehensive a record of any major historical event of its magnitude that we have.[6] Rather than serving as evidence of a digital dark age, the terrible events of September 2001 suggested that a golden age of memory had dawned.

This formative period for web archiving remains relevant to today's web archiving and digital preservation landscape. Between 1995 and 2001, the field witnessed the emergence of the Internet Archive, national library collection programs in Europe and North America, new international coordinating bodies, and a cultural consensus within the library and archives field that this work was both central and necessary. While this book is very much a story of developments in the Global North, primarily due to the high costs of these digital preservation programs and their accompanying infrastructure needs, the programs implemented in a small number of affluent countries came to have a global impact. Many of these programs were cemented by late 2001. If there had been any doubt, the collaborative mechanisms put in place after the 11 September 2001 attacks solidified policies and procedures for the future. The core programs that today form the foundation of the global web archiving landscape owe their origin stories to these critical years.

If there was a digital dark age for the 11 September attacks, it is measured in hours rather than days, months, or years. In 1996, our digital heritage was fragile and ephemeral. A digital dark age loomed. Five years later, a cadre of professionals moved into action within hours, actively preserving events as they happened. Between 1996 and 2001, dedicated information professionals—at the Internet Archive, libraries, and elsewhere—averted the digital dark age.

Dark Age to Golden Age: The Preservation Story

Averting the Digital Dark Age explores this shift from fears of digital loss to our current state of abundance. It does so by examining the intellectual ferment between the web's 1991 birth and the coming of age of web archiving programs in 2001, represented by both the 11 September attacks as well as the launch of the

Wayback Machine portal shortly afterwards. While in some cases the book looks forward a few years later to see the culmination of some processes, particularly around legal deposit and copyright, in the main it focuses on this critical period. In other words, if in 1996 and 1997 the affluent among Western society confronted the specter of a digital dark age, only four years later we had entered a period where we had the potential for a golden age of robust historical records. This was a rapid shift. Why does the web remember today? And what can we learn about this transition as we look ahead to future medium shifts?

Today, we grapple with the opposite problem of a digital dark age. What are the implications of all the remembering we do? Should people be beholden to comments they made ten years earlier, especially if made when a child? Do they even realize that the web remembers? Should data be allowed "to die"?[7] The web today is a mixture of fragility (the ever-present 404 error signaling that a page is missing) and permanence (the ability to pull up a deleted item from the hundreds of billions of websites preserved by the Internet Archive). The web does not have a built-in memory system, but libraries and other memory institutions around the world fill that function. What is collected and what is not—"archival silences"—are shaped by the historical processes that gave rise to this preservation infrastructure.

This unsettled situation has its roots in a series of decisions stretching back decades. On the one hand, the web inadvertently ended up as a fragile, ephemeral platform. The decisions that allowed it to scale rapidly across the world to become a "world" wide web also meant that links broke, servers disappeared, and a failure to pay a domain rental fee felled many a website. Realizing that, memory professionals pragmatically and presciently developed memory systems. As they were not integral to the web, many web users today (and over the last two decades) have been caught off guard by the limitations as well as the scope of digital memory.

In adopting the moniker of a "digital dark age," commentators and pundits were drawing on a long historical tradition of using this evocative understanding of the past. The framing of a digital dark age itself draws on an apocryphal understanding of the past. The period between the end of the Western Roman Empire and the Renaissance was erroneously understood in the past as a period of "darkness," a marked discontinuity between the Roman Empire and the early modern period. While the term has fallen out of historiographical use, there is still a popular understanding of the medieval era as a dark, superstitious, violent time (one rarely uses the adjective "medieval" as a compliment). Due to this legacy, the "dark ages" is a useful term for cultural commentators. As Matthew Gabriele and David M. Perry note, "the particular darkness of the Dark Ages suggests emptiness, a blank, almost limitless space into which we can place our modern preoccupations whether positive or negative."[8] For early Renaissance thinkers, the concept of a dark age was useful for drawing a distinction between the darkness of the recent past and the brightness of antiquity, which helpfully denied continuity between the Roman Empire and the Holy Roman Empire.[9] The term, and the historical implications stemming from it, have thus been useful to commentators for centuries. In any case, even during the period we dismiss as the Dark Ages, much of central Asia was undergoing an age of enlightenment, a flourishing of science, culture, and philosophy.[10]

Compounding this, record-keeping practices have always impacted our histories and led to inclusions and exclusions. In some ways, for example, we struggle to understand central Asia's age of enlightenment due to a lack of transcription and translation of critical manuscripts.[11] This period, an obvious counterweight to claims of a dark age of human history, is partially obscured by more recent decisions around records and now digitization. Selectivity around sources also influenced Western European historiography and helped contribute to the dark-age framing. Historian Patrick Geary notes that "what we think we know about the early Middle Ages is largely determined by what people of the early eleventh century wished themselves and their contemporaries to know about the past."[12] The Dark Ages are less a product of their own time—and more the outcome of decisions made by successors about what records to retain. History is often more about the stories that are told about the past and the records that are kept than about the past itself.

Archivists and record keepers are thus critical to the construction of the past and our history. If we have a digital dark age of the 1990s or 2000s in our future, it will owe less to the decisions of record keepers made at the time and more to our failed long-term commitment to stewarding this information over the decades and centuries to come. The real challenge of digital preservation is organizational, rather than changing formats or disks.

Discussions around the idea of a digital dark age, both in the media as well as among librarians, archivists, historians, and policy makers, set the stage for two main approaches to web preservation. The first, the Internet Archive, represented a private, nonprofit, technologist approach to the perpetual preservation of web content. Founded by technology entrepreneur Brewster Kahle in 1996, the Internet Archive responded both to prevailing cultural trends while also drawing on Kahle's earlier (and unique) experience across the digital libraries and information retrieval communities. Yet for the first decade (or longer) of its existence, many feared that the Internet Archive would disappear into the ether, plunging the web back into a digital dark age. With such an accumulation of information, this would be akin to losing a modern Library of Alexandria.

This private initiative was counterbalanced by a second approach, that of national library web archiving in several affluent countries. States can persist when private organizations fail. As early as 1994, the National Library of Canada piloted web-based archiving, followed in 1996 by large-scale projects carried out by the national libraries of Sweden and Australia. These programs offered the institutional stability that the Internet Archive appeared to lack, albeit at the cost of being less flexible due to government regulation and resource constraints. Together, the Internet Archive and national libraries formed an effective memory system.

These approaches complemented each other when it came to the long-term stewardship of digital material. The Internet Archive innovated, collected, inspired, and took on risk. Make no mistake: collecting the web at scale was risky business. Risk was present at every stage—of lawsuits, of breaking copyright law, of losing priceless and irreplaceable data. The Internet Archive could assume these risks in a way that an institution like the Library of Congress could not because of institutional contexts, bureaucracies, and a different risk appetite. Indeed, many national libraries continue to struggle today with collecting and access, often owing to internal risk calculations. The Internet Archive assumed a great deal of the world's web archiving risk. Several times, as we will see, legal commentators confidently assumed that they would be sued into oblivion. On the other hand, national libraries offer stability and sustainability. They may not have the innovative energy of an Internet Archive, but they should last longer. In any event, given the potential for state failure, it is good to have digital collections in both public and private hands.

Studying the Internet Archive, national libraries, and individual preservation initiatives in isolation does not do justice to the broad intellectual ferment that underpinned the digital transformation of libraries in the 1990s. To understand the rise of web archiving, it is essential to explore the intellectual conversations of the 1960s onwards, the rise of international networks, and those who brought the problem into public conversation. If in January 1996, the web was ephemeral, merely five years later, in 2001, the picture had become much more complicated. We live in a world today transformed by this shift. The digital records of elections, disasters (natural and human alike), pandemics, childhoods, cultural phenomena, memes, and so forth are all preserved to varying degrees.

Historicizing this question thus helps us to understand the limits of memory for the internet and web. Decisions from decades ago shape our record today. The web "remembers" through the work and policies of people and institutions around the world. Yet it does so imperfectly. Viktor Mayer-Schönberger argued in 2009 that "forgetting has become costly and difficult, while remembering is inexpensive and easy."[13] If this book shows anything, it is that the act of "remembering" the web is neither inexpensive nor easy. It requires continuous investment.

Much of the popular story of this problem and its "solution" revolves around the Internet Archive. Brewster Kahle, a Bay Area technology entrepreneur, who prior to the Internet Archive had codeveloped the Wide Area Information Server (WAIS) architecture, was at the end of a stint at America Online when he began to explore whether to establish an organization to preserve the internet. In early 1996, Kahle founded the nonprofit Internet Archive alongside the for-profit Alexa Internet corporation. That they were founded on the same day speaks to the meshing of commerce and altruism. Alexa would crawl data and use it to develop web navigation tools, whereas the Internet Archive would steward the data in perpetuity as a public good. In this arrangement lay the core of the Internet Archive's sustainability model. In some ways, the meshing of public interest and commerce can be seen in the Internet Archive's contemporary Archive-It service, which offers paid subscription services for institutions to carry out web archiving while drawing on Internet Archive infrastructure.

Yet the Internet Archive is not the whole story. In 1994, the National Library of Canada launched the Electronic Publications Pilot Project, or EPPP. This pilot investigated the technical and scholarly implications of harvesting selected scholarly journals and new media publications from the web. At a time when it was by no means assured that the web would be the dominant means to access the internet—competing protocols and platforms included Gopher, Archie, and WAIS—the EPPP laid a foundation for future web archiving. Its activities and 1996 final report influenced and inspired other web archiving projects.

In 1996, the National Library of Australia was motivated in part by the EPPP to begin harvesting websites, with an eye to curating a collection of culturally relevant sites for future research use. That same year, the Swedish Kungliga biblioteket (KB) launched its Kulturarw3 project. Yet the Swedes and Australians took very different approaches. Rather than curating a collection, Kulturarw3 instead sought to find every Swedish website. Arguing that the Canadians and Australians had been too narrow, the Swedes believed that harvesting web-based "ephemera" as well as formal publications would be more cost efficient and sustainable. If the real cost came in the time spent by the selector in choosing which webpages to archive, perhaps just letting the crawler loose on the web would be cheaper in the long run. From these early examples came other web archives, including the Library of Congress's MINERVA project in 2000. Debates over which approach to adopt spurred intellectual conversations around how to capture the web sustainably and effectively. Directing our gaze back to the debates around national libraries and the much broader intellectual ferment helps to broaden our understanding of why and how the web remembers.

Much of this coalesced in the aftermath of the 11 September 2001 terrorist attacks. Within the first hour, it was clear that the attacks would have profound political, social, and cultural importance. As Americans and others went online to check on one another's safety, to memorialize, vent, remember, debate, and collectively make sense of the events, national libraries, researchers, and the Internet Archive quickly preserved the events of the day and the months that followed. They created a community-based archive of the attacks and cemented the enduring significance of web archives and digital collecting amongst libraries and the public.

Despite this book's broad geographic scope, these preservation activities were mostly carried out in the Global North, defined in this book as inclusive of affluent countries that are primarily but not exclusively in the geographic north (for example, Australia is part of the Global North). The web grew throughout the period to become the predominant global network. As Gerard Goggin and Mark McLelland argue in their *Routledge Companion to Global Internet Histories*, we need to complicate our American and Eurocentric narrative of internet histories to think more broadly about the implications of how the internet was implemented and used in countries and regions around the world.[14] Parts of the world beyond the United States and Europe often used different systems in place of the web during much of the period discussed in this book, whether bulletin boards in Taiwan or email lists in Korea, which entailed different approaches to digital preservation. Indeed, the Taiwanese bulletin board system, PTT, has its historical record "curated by peer-appointed moderators . . . who exercise the right to periodically clean up a given board."[15] Such material can be crawled, but unlike the common standard of the web, it requires more bespoke technical approaches.[16]

Until recently, the infrastructure needed to preserve the web required a level of investment and resources found in few places apart from the private philanthropy and entrepreneurship of Kahle's Internet Archive or the digital preservation programs of a handful of affluent national libraries. Indeed, even today many national libraries rely on the Internet Archive to provide the core infrastructure of web archiving, finding the specialized staffing and infrastructure costs too onerous for their own operational capacity.

Yet the development of core web archiving infrastructure would allow for web archiving to expand beyond the Global North. Many Asian countries today have large national web archiving programs, as do several South American countries, although web archiving programs in Africa remain rare. Indeed, web archiving awareness and activity remains low across memory professionals even in relatively affluent African countries such as South Africa.[17] This movement outside the Global North is a recent one.

The interconnected nature of the web meant that the establishment of web archiving infrastructure in the Global North affected the whole world. As early as 2004, Peter Lor (then national librarian of the National Library of South Africa) and Johannes Britz of the University of Pretoria reflected on this imbalance. "There is little doubt that the national libraries and other responsible institutions in Africa and other regions of the developing world are not yet in a position to harvest and preserve the web sites emanating from their territories," Lor and Britz observed. Given South-North information flows, the coauthors noted that there were many legal and moral issues to explore. They suggested broad moral guidelines for practitioners to follow when archiving the Global South, including doing no harm, disclosing objectives and anticipated outcomes, focusing on reciprocity and equity, depositing data and publication, and—vexingly—considering the principles of informed consent and confidentiality.[18] The ways in which web archiving evolved, however, mean that informed consent is rarely possible. Crawlers cross borders and collect material at a rate that eludes human oversight.

The web grows exponentially, meaning that only a handful of institutions—perhaps a dozen—have the specialized training, expertise, and infrastructure to harvest web content at scale themselves. Web archiving, too, grows ever more complex. Yet the international nature of the web means that crawlers from the Global North routinely harvest material created in the developing world. They do so, however, through algorithms that largely represent Global North perspectives. The ensuing archival collection certainly reflects this bias. South-North information flows continue to shape global information ecosystems. In some ways this marks continuity with earlier national library collecting practices, from both foreign collecting to more reciprocal exchanges of government documents for purposes of safekeeping. The Library of Congress, growing in part out of international scientific collecting driven by the Smithsonian Institution in the mid-nineteenth century, amassed large international collections. Its international mandate would see (and sees today) selectors and curators amassing large collections of documents from across the world.[19] There is thus continuity in this, as national libraries today use web crawlers to select material, just as they also use purchasing agents or other selectors to curate physical material. Such activities are motivated by the desire to create global collections, as well as a nod toward the value of distributed preservation. Accordingly, even if much of the scope of this book is limited to the development of this infrastructure in the Global North, the impact continues to be global.

The core contribution of *Averting the Digital Dark Age* focuses on the period from 1995 to 2001, although there is essential context before and after that period that informs this narrative. Much of the first chapter reaches back decades to understand the broader history of what we today call digital preservation. It was only in 1995, however, that the explosion of user-generated content made the prospect of a digital dark age chilling beyond the narrow scope of corporate and institutional records. Digital preservation escaped the staid world of record and archival management and became a pressing social problem. It is one thing to preserve the records of a Fortune 500 company. Records managers can help there. It is another to ensure that your own website, newsgroup postings, and online relationships and communities have a life beyond that of a webmaster's whim or ability to pay a bill. The digital dark age and the broad problem of digital preservation became a matter of public importance in 1995.

Few histories have straightforward end dates. As I note in my conclusion, digital preservation requires continual attention if it is to be sustainable. Indeed, digital preservation and web archiving remain active areas of research and interest. Technical practitioners discuss best practices for preserving archived material, have conversations about how to expand the curatorial scope to include the voices of marginalized or otherwise underrepresented people, and continually explore opportunities to improve the capture of dynamic events such as protests or conflicts. Can the story of *Averting the Digital Dark Age* end in 2001 when the field continues to evolve and flourish? Is it hubris to say that the digital dark age has been averted? The conversations around the preservation of the 11 September 2001 terrorist attacks marked a moment when the social debates (Should we preserve this content?) gave way to technical discussions (How can we ensure the fidelity of what we are capturing?). Yet, conscious that an abrupt 2001 ending would unduly cut short my narrative, I carry some stories forward. There is a vast technical literature on this topic, complemented by only a small body of work from the humanities or social science. This reflects the move away from fundamental philosophical questions about whether we should avert a digital dark age and toward the question of how to do so.

The Scholarly Conversation and Structure of This Book

Averting the Digital Dark Age explores how the web stopped forgetting and came to remember. In doing so, it intersects with robust scholarship on web archives as well as libraries and memory more generally. Scholars have explored the theoretical and applied impacts of contemporary web archives, most notably the path-breaking work of Niels Brügger.[20] There is also a growing literature on the use of these collections, including research case studies, ethical explorations, technical refinements to crawling or analysis, and beyond.[21] It is also part of a broader field of digital preservation, most accessibly and thoughtfully found in work by Trevor Owens.[22]

Averting the Digital Dark Age is not a technical guide. While some technical details will be discussed where appropriate, the book emphasizes the social and organizational infrastructure and apparatus that made web archiving possible at scale. Much of the extant literature on web archiving is technical, and indeed, much of the gray material around web archiving stems from the technical conversations between experts on how to carry out this form of archiving. Throughout this book, however, where technological development played a major role in the programs discussed, it will be appropriately centered.

Given the technical bent of today's web archiving literature, it is perhaps unsurprising that the history of the Internet Archive and web preservation more generally is reduced to several rote paragraphs in most works (including my own earlier work). In these abbreviated treatments, we generally learn that the Internet Archive was founded in April 1996, with some other national library programs beginning shortly thereafter. Few details and context are provided. In general, the narrative of the digital dark age and its "solution" focuses nearly exclusively on Kahle and the Internet Archive, with perhaps a few nods to a handful of other programs in a sentence or two. This is not to underplay the Internet Archive's significance. Despite the discomfort in contemporary historiography toward hagiography, web archiving's history is intertwined with Kahle's vision and efforts, as enacted by the Internet Archive. Yet a disproportionate focus on the Internet Archive neglects the broader cultural moment that gave rise to sustainable web archives around the world and made the Internet Archive's success possible.[23]

Beyond the web archiving field, there is a more general literature on digital memory. There web archiving is often reduced to a short overview. Richard Ovenden, for example, notes that in a hundred years, scholars will look to our digital records as sources. In his words, "there is still time for libraries and archives to take control of these digital bodies of knowledge in the early twenty-first century, to preserve this knowledge from attack, and in so doing, to protect society itself."[24] He understandably worries about the Internet Archive's sustainability. Ovenden is correct that there is more work to be done around institutional digital preservation and web archiving, but arguably institutions both recognized and "took control" of this problem from the late 1990s onwards. Other works, such as the monumental, edited collection *Information: A History*, only briefly discuss the Internet Archive (understandable given the volume's broad mandate): highlighting its importance for preserving information but quickly brushing past its origin story with a few nods toward the Long Now Foundation and the web's lack of an intrinsic memory function.[25]

While this book is in conversation with the fields of new media history and media archeology, it is at its core a work of historical scholarship. It aims to provide context for media archeologists, but it is not itself a work of media archeology. Informed in part by my grappling with the history of digital preservation, my emphasis is on the representations found within media objects. I am less interested in the physicality of a hard drive or HTML encoding but rather what the represented data, placed in context, tells us about the broader social world.[26] Media archeology is in part a conscious rejection of histories that are a "telling of the histories of technologies from past to present."[27] This point is well taken: few historians today would write Whiggish histories of progress, and I am not doing so either. While this book has much to learn from these theoretical approaches—especially around the pitfalls of avoiding nostalgia and adopting these theories as an "analytical tool," per Wolfgang Ernst—it is at its core a work of traditional historical scholarship.[28]

For this project, I am especially indebted to those studies that situate earlier media technologies or conceptions into broader conceptual frameworks. Ideas of hypertextual communication and the augmentation of human memory often harken back to ideas such as Vannevar Bush's Memex (discussed at length in chapter 1). Wendy Hui Kyong Chun's work suggests that Bush believed that he could break history's discontinuity—a process "due to a historical accident, to our inability to adequately consult the human record, to human fallibility."[29] Just as the Internet Archive's Wayback Machine seeks to fix the web and provide a memory, Bush's conceptual 1945 Memex aimed to provide an all-encompassing view of human memory through analog microfilm technology. Chun's attention to terminology and concepts is helpful, making us think about what we mean by putting information into "memory"—usually we *recall* from memory, not consciously put things into it.

The work of Matthew G. Kirschenbaum, too, underscores the complexity that underlies digital files and objects. By looking both at the big picture as well as the "bitstream" of a file, we can understand the longer sweep of digital preservation. In his 2021 *Bitstreams*, Kirschenbaum gets into the weeds: what does it mean to study, say, a manuscript that was drafted in Word, commented on in iCloud, laid out in Adobe, with versions stored on USB drives, and with sequential and often obscure file names?[30] Any deep dive into the Internet Archive reveals the necessity of grappling with these details, and thinking forthrightly about what this kind of scholarship means.[31] Crucially, Kirschenbaum is conversant with the sweep of archival conversations around digital records stretching back into the 1960s. His observation that the "shift from *an* archives (as a place) or even *the* archive (as a trope) to *archiving* as an active and ongoing process" was anticipated by archival theorist Terry Cook and requires a different paradigm around the preservation of knowledge. Archiving was previously seen as passive but rather now "is best understood as a continually active *process*, requiring ongoing care, attention, and maintenance to ensure that systems remain secure, software stays up to date, connections don't deteriorate, and bits don't rot."[32] Indeed, Kirschenbaum's earlier *Mechanisms* underscored the challenge of digital preservation as "while massively technical to be sure, are also ultimately—and profoundly—social . . . effective preservation must rest in large measure on the cultivation of new social practices to attend our new media."[33]

There is some irony in that memory institutions themselves are often difficult objects of historical study. The archives of libraries and archives are themselves limited. As many institutions confront accession backlogs for external content, their more recent institutional records remain inaccessible. Given the relatively recent period of this book, this is an especially acute problem. Fortunately, libraries and archives are avid creators of gray literature (position papers, pilot projects, task force reports), all of which are critical to understand their thinking. In the case of the Internet Archive, and Kahle specifically, much of his private correspondence and files have been fully scanned. Their candor and comprehensiveness are testament to Kahle's commitment to open knowledge. These documents are supplemented by a series of interviews I carried out with key individuals from national libraries, universities, and the Internet Archive.[34] Margaret Hedstrom also generously provided me with material from her own holdings.

Given the book's scope, I made difficult choices around which institutions and programs to focus on, generally preferring to look at the early adopters. This means that some now-prominent web archiving programs are rarely discussed. One of the most obvious examples of this is the omission of the development of big national library programs such as those of the British Library, the Bibliothèque nationale de France, or the Danish national library. My book is rather the story of the pioneering institutions that laid the groundwork for these larger programs to subsequently thrive in the 2000s and 2010s.

The book advances its argument through a series of five interconnected chapters, proceeding roughly chronologically. The exact disentangling of the cultural moment in 1996 presented challenges in how to order the chapters. The Internet Archive is the focus of chapter 3 and national libraries in chapter 4, but the two chapters complement each other and could be read in either order. By separating them, I am conscious of drawing too firm a division between their approaches. However, their separate institutional lineages and the crucial role played by Kahle lend themselves to separate yet related treatments.

Accordingly, the book opens with chapter 1, "Why the Web Could Be Saved: From Machine-Readable Records to Digital Preservation." This chapter provides the context behind the preservation of a society's memory in archives and libraries, underscoring the importance of digital preservation. It then discusses the long sweep of attempts to organize the world's information, from 1890s global catalogs to the work of Vannevar Bush to the web itself. Finally, echoing many other voices in this field, the chapter argues that we need to understand digital preservation as an organizational rather than a technical challenge. All these disparate forces and factors came together with the web and its preservation.

With the need for web archiving established, chapter 2 is "From Dark Age to Golden Age? The Digital Preservation Moment." It explores the rise of the idea of a digital dark age and the overall shift toward changing our understanding of electronic records and their implications. It does so by following a series of individuals and organizations including information scholars such as Margaret Hedstrom, Microsoft executive Nathan Myhrvold, science fiction author Bruce Sterling, and the Long Now Foundation, who all helped make the idea of digital preservation more accessible to a general audience. In only a few years, the importance of preserving digital information became accepted within information professions as well as cultural conversations more broadly. If in 1991 the web was ephemeral, by 1997 it faced the prospect of beginning to remember. The next step was to build the necessary infrastructure to make these visions possible.

Chapter 3, "Building the Universal Library: The Internet Archive," pivots to explore the Internet Archive's origins, as well as the background of its founder Brewster Kahle. It explores the many factors that gave rise to the institution, as well as exploring how a small startup could grow to define and create the international web archiving landscape. Ultimately this was possible because of the intersection of the cultural forces discussed in chapter 2. They had provided an audience and a foundation of a ready audience convinced of the importance of preservation, transforming what might have been a hobby project into one that could shape the global digital landscape.

Chapter 4, "From Selective to Comprehensive: National Libraries and Early Web Preservation," explores how national libraries adapted their policies and approaches to the digital age. It looks at Canada, Sweden, Australia, and the United States to explore how the earliest web archiving operations began as well as how they debated amongst themselves as to whether a selective or comprehensive approach would be most effective. These were debates about whether the mass collection of information would lead to large, low-quality, and ultimately less useful, collections as well as what it would mean for the role of a web archivist as a creator.

Finally, chapter 5, "Archiving Disaster: The Case of 11 September 2001," brings together the currents discussed throughout this book to see how the web archiving community—both the Internet Archive and national libraries—responded to the consequential events of that day. In the immediate hours that followed the terrorist attacks across the eastern seaboard of the United States, it was apparent that the events would represent a historical political, social, and cultural moment. Archivists and librarians moved into action, creating a robust historical record, complementing it with all sorts of digital collecting activities. The digital dark age of 11 September 2001 would be measured not in months but hours. I then conclude the book with reflections on how historical lessons can inform contemporary discussions around privacy and the right to be forgotten today. What have we gained and lost?

Much of this book was researched and written during the COVID-19 pandemic. For many white-collar workers, one enduring image of the pandemic might be a screen: a place to work, connect with extended families, nourish friendships, and beyond. Future generations trying to make sense of COVID-19 will do much of this through electronic records, from government dashboards, social media accounts, and disinformation around the virus and vaccines alike. In their research, they will be the unwitting beneficiaries of the legacies discussed in this book. The web now remembers, for better or for worse. By looking back at how the first professionals and activists responded to the rise of the web and the accessibility of the internet, we can better document and preserve our world today as well.

CHAPTER ONE

Why the Web Could Be Saved

From Machine-Readable Records to Digital Preservation

Without intervention, a conversation, an event, an action—anything that happens, really—is eventually (and usually quickly) forgotten. Perhaps the event lives on for a bit longer in our human memory, before being lost through aging or death. Very occasionally, an event is documented as part of an official record. Or, also rarely, it is captured on a scrap of paper, in a diary, or by somebody's iPhone recording video. Yet, unless this record is subsequently committed to long-term storage, it too will almost inevitably—eventually, inexorably—disappear.

Archives and libraries must think on long time scales to preserve and steward information. Since the advent of the modern historical profession in the late nineteenth century, historians have drawn heavily on primary documents—documents produced by the object of study or at the time of the event that the historian is studying—that are stored in and by archives. While source bases have more recently expanded to include oral histories, folklore, Indigenous stories, and beyond, there is still a professional bias toward the written word. Archives have never been complete or representative: it was good to be rich, white, and powerful if you wanted to be remembered within institutional structures. The poor, young, racialized, and working-class were often understood only through sources that incidentally captured their experiences. Moving into the 1980s, while archivists and historians were aware of the limitations of their archives, there was a common understanding of how documents and paper could be accessioned into an archive, described, and preserved and made accessible for future generations.

Then the desktop computer changed everything. The increasing volume of electronic records, and later networked ones, threatened to upset traditional archival workflows and thus our society's approach to long-term textual memory. While archivists were already dealing with the explosion of paper records that accompanied the growing bureaucratic state and modern office culture, electronic records seemed to exist on an entirely larger scale of information. These "born-digital" records required specialized expertise to ingest, maintain, and provide access to. A paper document, properly preserved in a climate-controlled room in an acid-free archival box, will last a long time. An electronic document, on the other hand, resided both on fragile media *and* prompted worries that there would be no software with which to read. One cannot squint at a digital record and see the 1s and 0s. Even if they could, the 1s and 0s are meaningless absent their translation into intelligible information. By the 1980s, this was compounded by the rise of networked documents and information, a trend exacerbated by the advent of the web in the early to mid-1990s.

This chapter provides the context of digital preservation before the web and during its earliest days. This lets us understand how, despite the lack of an innate memory function—by default, the web tends to forget—the web could actually be saved. This was thanks to the efforts of an earlier generation of digital preservationists who laid a foundation for the conversation that would follow. Beginning with an exploration of why we need to understand archives, the chapter then discusses the early history of networked conversation and hypertext. It then shifts into the broader field of digital preservation, with emphasis on why organization and personnel challenges are far more significant than the technical bits and bytes of a born-digital file. Hardware is not the crux of the problem: policy, procedures, and organization are.

Why the Histories of Archives Matter

Understanding how a society constructs and stewards its archives and archivists is critical to the comprehension of its historical scholarship. Archival silences are shaped both by the context in which records are preserved and the way historians subsequently use them. Historians need to better understand the archives that they use to write their histories. Archives arguably shape our understanding of the past just as much as historians do. In a call for more archival histories, Elizabeth Yale argues that historians need to "understand the histories that have shaped [archives]: these histories constrain the kinds of stories that can be written from any particular archive."[1] Francis X. Blouin Jr. and William G. Rosenberg echo this in their work on the complicated relationship between archivists and historians, noting that our need to understand archives as active agents moves "archives from a place of enquiry to a subject of enquiry, from a locus for research to an object of research."[2] While historians are vaguely cognizant of the role that archives play in shaping knowledge, there is still a broad tendency amongst historians to treat archives, as Alexandra Walsham has argued, "as neutral and unproblematic reservoirs of historical fact."[3] To outsiders, historians' lack of engagement with archives may seem surprising—yet Walsham is right. Many historians have not, in general (there are many field-specific exceptions such as Indigenous histories or postmodernist historians), thought enough about archives: their origins, histories, and the theoretical conversations that underpin them.[4]

This is unfortunate. Historians need to be more cognizant of archival scholarship and to be conscious of archives and archivists as active institutions and professionals who actively shape historical knowledge. Archivists are often wholly absent from historical scholarship save for words of thanks in acknowledgments. As professions, historian and archivist have sufficiently diverged that Blouin and Rosenberg make the provocative but no less accurate claim that "any visit by a historian to an archival institution is now an exercise in interdisciplinarity."[5] By being conscious of this, historians are more self-reflective in their work.

For the decisions about what to preserve, accession, and describe—and which documents to *not* preserve—has dramatic impact on how we understand a historical era. The long-standing archival appraisal process transforms the way in which information is preserved and used.[6] This is as true for historians studying the seventeenth century as it is for those of us studying the more recent past.[7] As archival theorist Terry Cook has argued, archivists "co-create the archive" through appraisal. They do so by actively deciding what, from the scant documentary record, might be preserved, 1% to 5% of which they have in turn been tasked to appraise and catalog.[8] To Cook, "a major act of determining historical meaning—perhaps the major act—occurs not when the historian opens the box, but when the archivist fills the box."[9] While proponents of "objective history" might believe that a historian could visit an archive and examine an unmediated past, understanding how the historical record is created is key for responsible historical scholarship.

Traditionally, historians have dated their profession's engagement with archives to Leopold von Ranke and his late nineteenth-century establishment of modern, document-fixated professional history. Historians, however, had been using archives for centuries before von Ranke. Archives have a long history of their own too, independent of historians. Markus Friedrich has argued that while the impulse to centralize documents stems from as early as the late twelfth century (Richard Lionheart's defeat of Philip II and the capture of his royal documents at the 1194 Battle of Fréteval led to centralized record keeping), by the early modern period there was an emerging consensus around the importance of protecting and preserving documents.[10] By the early eighteenth century, it seemed impossible to some historians (even before von Ranke) that one could carry out historical work without archives. Philipp Julius Rehtmeyer wrote in 1707 that such an undertaking "would be just like someone who set out to work on a religious book without first diligently gathering readings from the Bible."[11] Historians and archivists thus have a long-intermeshed history, stretching back centuries.

Yet history is never uncritically mediated through archives. Archives were and are, by necessity and design, selective institutions, shaping the record that early historians and contemporary historians alike used and use.[12] By the mid-nineteenth century, when von Ranke pioneered his vision of scientifically objective history, holding that through archives a historian could understand the past "wie es eigentlich gewesen ist," or roughly, how it actually was, he was drawing on a long lineage of archival historians. Yet von Ranke's emphasis on letting evidence "speak for itself" was a strong, empiricist statement that shaped historical research.[13]

This emphasis on impartiality was mirrored within the archival profession, representing the then-in-vogue positivist strain in Western intellectual culture. Sir Hillary Jenkinson's 1922 *Manual of Archival Administration* articulated a straightforward role for archivists as archival custodians. This vision was of impartial professionals stewarding documents that were deemed authentic by virtue of their relationship to the record producer. Historians could thus then use these documents to see the past "as it was," unmediated by archivists (in this theory) who were simply passing knowledge on to be preserved.[14] The bulk of archival information always made this imaginary role an impossible one, a fact compounded by the rapidly growing volume of archival information throughout the rest of the twentieth century.

Archivists make decisions that unavoidably shape the historical record. If in 1922 it was possible to pretend that archivists were just passing along unmediated information, the sheer volume of information a few decades later made this untenable. At least partially in response to ever-growing archival "bulk," by the 1970s archival theory moved into a "post-custodialist" phase. Articulated by archival theorist F. Gerald Ham, post-custodialism posited that archivists were "activist archivists." This was not a political position but rather a recognition of the role of archivists in shaping and influencing the historical record. Ham argued that archivists could no longer be passive custodians in an "age of abundance."[15] Indeed, in 1974, Ham put this clearly: "the archivist must realize that he can no longer abdicate his role in this demanding intellectual process of documenting culture . . . the scope, quality, and direction of research in an open-ended future depends upon the soundness of his judgment."[16]

Yet the 1970s witnessed professional estrangement between historians and archivists—just at the moment when conversation and dialogue was especially critical. The reasons for this are complicated, linked to growing professionalism amongst archivists as well as the growing scope of historical inquiry. The social (and later cultural) turn had moved historians away from traditional archives, further removing some common touchpoints. If a political historian used a fairly predictable set of official documents, a cultural or social historian might be more interested in advertisements or marginalia—increasingly idiosyncratic source bases that were less consistent than those used by their counterparts in political or diplomatic history.

This estrangement led to the present situation, where we can perhaps even think of historians using archives as engaging in interdisciplinary engagement, as their shared professional origins are now decades in the past. Edward Higgs, a historian who then worked at the United Kingdom's Public Record Office in the late 1970s to the early 1990s, witnessed this divorce between historians and archivists firsthand. As he recalled, "there was a great deal of hostility from the developing archival profession, many of whom wanted to see themselves as information scientists rather than the old-fashioned custodians of the 'handmaidens of the historical profession.'"[17] Yet many historians implicitly still thought of their archives along those lines. Archives need to be studied, not just as neutral repositories but as sites where history and meaning are actively constructed. "Archives are not storage facilities or receptacles that simply accumulate documents," Friedrich argues. Archives "are the sum of activities and actions."[18]

Electronic records exacerbated these trends. Technology threatened established approaches to archival selection. While it is impossible to keep *everything* that is produced, what if web archivists could keep everything that they had at least found in the course of their scoping? Would this fundamentally transform the relationship between archivist and document? What if selection was handed over to an algorithm? What role would the web archivist continue to play as an active agent?[19] We must also keep in mind that selecting something often means *not* selecting something else, even if by virtue of inattention or in other cases due to budgetary restrictions. This makes the remaining documents even more significant in their survival. In other words, as more things are kept, the relative salience of the average individual document diminishes. Questions that needed to be explored by archivists and librarians included: should websites be preserved? What of those created by governments? Or those by everyday people? Were webpages transient ephemeral documents, akin to email (until 1989 in the United States, emails were only routinely preserved if they were printed).[20] Or were these modern publications, a written record of our time? The pressing nature of digital preservation meant that archivists and librarians would not have the benefit of time as they considered these questions.

Digital Preservation: A Short Overview

The digital dark age problem is an intuitive one. We can all see the longevity of paper, living a long life on our bookshelves, in our shoeboxes, and attics. Conversely, we can see how quickly files and software disappear. Indeed, early internet users were often the first to truly realize just how transient electronic records were. Terry Kuny, who would later help popularize the idea of a "digital dark age," recalled watching the Soviet invasion of Latvia in 1991 playing out over the BALT-L BitNet listserv. To his surprise, he realized how ephemeral all of this information was: "I was discovering that no one was keeping track of these records. I ended up sending a ream of paper. The old tractor-feed paper . . . sending it off to I think was some Russian study group somewhere in southern Ontario as an archive, thinking that they might want [it]. I bound it all up and thought, well, this might be a really interesting record."[21] For those like Kuny, this omission inspired action as the scale of potential loss became clear. These problems would accelerate as increasing amounts of society's memory ended up on third-party platforms, "effectively outsourcing that memory to the major technology companies that now control the internet."[22]

Digital destruction presents an ever-present risk. Of course, "analog" information is also at risk. Most paper also disappears, not to mention television recordings, radio broadcasting, and ephemeral telephone conversations. Yet, with the shift to digital information, much more information than ever before will end up being saved in absolute terms. As people move their lives onto third-party digital platforms, it is thus crucial to steward it as best as our society can. Primary documents of the BBS era or PLATO are largely gone, for example. Imagine if the web used by billions today were to meet a similar fate.

Transmitting and preserving information from past to present is difficult. Paper rots. Stone crumbles. Language changes. Memories, passed down through generations, become fuzzy or lost depending on memory traditions. Forgetting is the default. Information, as information scholar Michael Buckland explains, is both "what we infer from gestures, language, texts, and other objects," but also what we can conveniently refer to as documents: "material forms of communication—bits, books, and other kinds of physical messages and records."[23] Through preservation, from cave paintings to telegrams to photocopying to digitization, these documents can travel across space as well as time, forming our historical record. Preservation, arrangement, and description by archivists transforms what would be a disconnected mess of information into something usable.[24] Digital information has compounded this long-standing challenge, with its attendant vast increase in sheer information. Vexingly, of course, digital information is unreadable without interpretation. As Buckland observes, "some kind of special rendering or visualization is necessary even for plain text."[25]

In general, there are three main approaches to preserving digital information.[26] They can be used alone or in combination. First, migration involves copying information to new media. As an example, data could be transferred from an external disk to a hard drive. One could then continue to migrate the documents onto modern formats again every few years afterwards. In practice, as storage capacity increases, migration takes place when a data center replaces, for example, 4TB hard drives with 8TB hard drives. The physical space on server racks is a scarcer commodity than the drives themselves, so the new drives bring both new life span and renewed infrastructure. Second, standardization involves moving material into a smaller number of longer-term file formats: migrating proprietary word processor files to PDF, for example, or standardizing a host of idiosyncratic image formats to JPEG. There are trade-offs with this approach, as some of the original characteristics of the documents can be lost. Yet it helps to preserve content without needing to steward legacy software. Third, emulation is a solution that sees older operating systems and environments virtually generated. For example, the Internet Archive uses emulation extensively on its site to allow people to play early video games. A virtual emulation environment starts, providing a user access to a virtual MS-DOS prompt from two or three decades ago.

The scope and scale of these broad approaches underscore the difficulty facing the digital preservation field. The landscape behind these strategies has also evolved considerably. In the 1990s and early 2000s, for example, physical media was a major issue. A 2003 anecdote from historian Roy Rosenzweig—who we will return to in chapter 2—underscores the seemingly bewildering pace of change confronting computer users: "At the moment [in 2003], for example, I am trying to 'migrate' my old 5¼″ floppies to something more modern. I am using the last machine that I have (and I oversee a lab and set of offices with twenty or twenty-five computers) that has a 5″ drive that still works. And even that requires copying from 5″ drive to 3″ floppy. Then copying to a Windows machine, then burning a CD, and then copying that to Mac. And then I need translation software to read the old WordStar files."[27] This was a personal experience that was more broadly generalizable.

In a 1992 report for the Commission on Preservation and Access, computer scientist Michael Lesk noted the various reasons that new media were a particular problem for libraries and archives. While deterioration was a standard concern of libraries and archives (archives closely monitor humidity levels to ensure the longevity of archival holdings), the additional digital factors of technological obsolescence, proprietary vendors going out of business, and the intermingling of "format, software and hardware" made for a very worrying landscape. To Lesk, the "survival of digital information [did] not depend upon the permanence of a particular object, but upon widespread distribution of the information, and regular refreshing of it onto new technology."[28]

Lesk's report resonates today. Dramatic transformation was necessary for libraries and archives. Rather than just investing in an object's one-time preservation and providing an appropriate environment, institutions would instead need to invest in a long-term strategy of continually copying and maintaining information.[29] This is not to set up another false binary. Preservation has never been "set it and forget it," whether it has involved a computer disk or a box of paper documents. Yet digital preservation adds another level of difficulty and requires more (and different) resourcing. The 1996 report of the Task Force on Archiving of Digital Information echoed this when it made the case for "migration as an essential function of digital archives" and looked forward to the prospect of the web as unleashing "the production and distribution of digital information."[30]

Given the preservation benefits of copying and distributing information, at first glance distributed networks like the internet and web might seem to be ideal places to preserve information. A global interconnected network can both share resources and distribute copies widely. For books, copy censuses began to be carried out at the turn of the twentieth century, seeking to understand the surviving numbers of rare books in global collections.[31] Indeed, this was the model that underpinned Nelson's "Project Xanadu" (discussed later in this chapter). Some groundbreaking preservation programs have indeed leveraged the widely distributed and networked nature of the internet, such as the Lots Of Copies Keep Stuff Safe (LOCKSS) project that has become an open-source application and network to facilitate the peer-to-peer preservation of digital material. With LOCKSS, copies are distributed across the network. Peers then periodically compare their documents with each other to ensure the integrity of their data. This approach avoids the vulnerability of a central registry and represents the potential of what the internet can achieve when it comes to preservation.[32] Yet LOCKSS would begin only in 1999, and it requires that libraries and communities have expertise and active engagement to join. By default, the internet forgets. This problem was compounded as the theoretical peer-to-peer nature of the internet gave way to one dominated by a much smaller number of critical servers.[33]

Potential solutions to these problems of ephemerality would come from the libraries and archives sector. Tasked with stewarding intellectual resources, both those produced by society as well as university researchers, the core ideas that would inspire long-term web preservation began to percolate within research libraries in the late 1980s and early 1990s. Yet they did so inspired by a long lineage of practitioners who had been exploring the problem of preserving digital records for decades.

The Long History of Digital Preservation

While "digital preservation" is a relatively recent term, the practice of needing to accession and preserve digital information stretches back into the middle of the twentieth century. The terminology evolved over this period: from the early 1960s into the mid-1980s, the field tended to be described as "machine-readable records" (such as stacks of punch cards or records encoded onto tapes), giving way by the middle of the 1980s to the more expansive vernacular of "electronic records."[34] Only by the 1990s would the term "digital preservation" come into vogue as a descriptive phrase, describing what had been happening for decades.

The popular history of digital preservation often involves the 1960 decennial census in the United States: the first census scanned and processed onto magnetic tape using the Film Optical Sensing Device for Input to Computers system (processing punch cards for tabulating had a much longer history, dating back to 1890).[35] The 1960 census myth holds that the National Archives recognized the importance of the tapes for historical preservation only by the mid-1970s, whereby at that point the physical tape drives were obsolete and unreadable.[36] The most extreme and popularly digestible account, found in the 1985 *Report on the Committee on the Records of Government*, claimed that by "the mid 1970s, when computer tapes for the 1960 census came to the attention of archivists, there remained only two machines capable of reading them. One was already in the Smithsonian. The other was in Japan!"[37] While the original census response cards were intact, this raised fears that the machine-generated derivative data around the census might have been lost because there was not a machine to connect these tapes to. The idea of needing to break a machine out of the Smithsonian or to fly to Japan just to read information confirmed the worst fears of digital loss.

The reality was less alarming, underscoring the degree to which at least some archivists and memory professionals were equipped to handle these new challenges brought by the digital age. In 1975 and 1976, staff did encounter difficulties with reading the census tapes but succeeded in *migrating* the information to modern tape drives. Ultimately, only two tapes out of 7,297 were unable to be recovered—and that was because they were lost ("inadequate inventory control") rather than technologically obsolete.[38]

The myth of the 1960 US census is thus a useful introduction to the broader field of digital preservation as it reveals several critical dimensions. Hardware obsolescence matters, but more important are policy, procedures, and the organizational attention paid toward information preservation. Like all good myths, it also contains a few grains of truth. While archival organizations had been preserving computer punch cards for decades—in 1939 punch cards were added to the Record Disposition Act in the United States—the 1960 census had been the focus of the National Archives' "first substantial efforts to preserve electronic information."[39] Dovetailing with contemporaneous efforts by quantitative historians who were meeting with archivists around the retention of data files, organizations were beginning to work on this problem. Meyer H. Fishbein, director of the Records Appraisal Division at the US National Archives, read a paper in September 1970 at the Society of American Archivists annual meeting, arguing that the moment for action was upon the profession. Unless archivists "fulfill[ed] our responsibilities, about one million reels of tape in the Federal Government and more elsewhere will be erased without any archival judgments on the continuing value of the information they store."[40] Yet Fishbein was hopeful for a future where finding aids would be made machine-readable and where archives would "communicate and transfer their information to researchers by remote control circuits."[41]

The pathway toward this vision would be a long one. The Data Archives staff at the US National Archives was established in 1968, spurred on by concerns raised by the Social Science Research Council's Committee on the Preservation and Use of Economic Data.[42] The first electronic records were subsequently accessioned in April 1970. By 1977, the Center for Machine-Readable Records was established, becoming the Center for Electronic Records in 1988.[43] The story was more complicated in other jurisdictions, absent the unifying force of IBM platforms within the US government or the American approach, which saw government publications as public domain. In the United Kingdom, where records are Crown rather than public property, the landscape meant that central archival bodies tended to advise rather than exercise more direct control.[44] A survey of these histories, however, underscores the organizational and policy-based challenges of digital preservation *rather* than a challenge rooted primarily in hardware or software.

This is not a novel argument. Despite a perception that digital preservation is a technical challenge, practitioners and scholars underscore the social or organizational challenges.[45] In his sweeping overview of digital preservation, Trevor

Owens argues that "preservation is the result of ongoing work of people and commitments of resources," a job that is never truly complete. Per Owens: "nothing *has been* preserved, there are only things *being* preserved."[46] This is true of all collections, of course, but the nature of the digital—1s and 0s preserved on storage media—means that it truly requires a sustained investment and attention if the information is to persist.

While centers of excellence were established at large institutions like the National Archives, the archival profession *writ large* was slower to adapt. Across North America, digital preservation was the purview of a small number of government organizations—the National Archives of the United States, a few states (Kentucky and New York), the National Archives of Canada, and a handful of academic initiatives—reflecting both the slow pace of adoption by archivists as well as the considerable overhead costs required to get into this field.[47] In May 1983, archivist Richard Kesner gave a keynote to the Annual Meeting of the Association of Canadian Archivists, warning that if archivists did not adapt to "automated information systems . . . before too long we will be relegated to the antiquarian curatorial role that we have heretofore rejected as a misplaced 'popular' notion of what an archivist does for society."[48] Seven years later in 1990, when staff at the National Historical Publications and Records Commission reported on the field, they noted that "managing information in electronic form to ensure its availability for future use by a broad spectrum of users . . . is the most significant and difficult challenge currently confronting the archival community."[49] The case was not too different by 1994.[50] Archivists understood that there was a challenge facing the profession—but this awareness had not yet been matched by concrete action to redevelop their approach.

The critical issues of digital preservation were (and are) not technical ones but rather organizational ones. As Lesk remarked in a 2014 retrospective, the "greatest danger to digital materials is that we forget the meaning of them . . . Keeping knowledge, rather than objects, is an organizational problem."[51] Sustainable digital preservation requires attention to the "digital preservation triad": the interplay between management, technology, and content.[52] It is complicated. There is the very real continued maintenance of integrity on disk, ensuring that you do not end up with the debt of an obsolete physical format or software encoding (requiring complicated decisions around whether to migrate file formats to contemporary standards or to develop emulation formats). This is all compounded by the exponential growth of digital information. When it comes to government data, for example, "the amount and complexity of data produced threatens to overwhelm the ability of the government to preserve it."[53]

Organizational challenges are significant. Despite the long history of digital preservation stretching back into the 1960s, by the early to mid-1990s the field was still nascent. In 1994, archival studies professor Richard Cox looked back to the 1960s and forward to his present day, concluding that the "archival profession in the United States [had] not done well in structuring itself to manage electronic records."[54] While alarm bells were beginning to be rung by scholars in journals such as *Archivaria* and *American Archivist*, Cox argued that there was no consensus on archival theory and electronic records, almost no job advertisements (nineteen jobs advertised between 1976 to 1990), and effectively no professional training or curriculum to deal with these issues.[55] Cox was blunt about the state of affairs: "it appears that the archival perspective has largely been blocked out of being able to deal with the major instruments in the creation and management of that information, the computer and computer software."[56]

The archival profession is a small one, and the world of electronic records or later digital preservation was far smaller. A small number of dedicated practitioners and archival theorists thus had the ability to significantly shape the evolution of the field. Margaret Hedstrom, who we will learn more about in the next chapter, was part of this bridge. Hedstrom was part of the first generation of electronic records specialists and subsequently was then present at web archiving's inception. These histories were thus part of the original visions of web archiving writ large. Perhaps uniquely among many of these voices, in her 1984 book *Archives & Manuscripts: Machine-Readable Records*, Hedstrom articulated not only the challenge (fragility and the requirement to decipher via hardware and software) but the opportunities presented by digital records. Archives faced storage challenges, whereas digital records were "very compact, readily accessible, and easily manipulated." Moreover, such information could be "rearranged, aggregated, compared, and subjected to statistical tests without the laborious tasks of data collection, coding, and data entry by the researcher."[57]

The difference between the conversation around electronic records in the 1980s and the conversations that followed was that it was still largely institutionally focused. Even historians exploring the problem tended to think of electronic records in relationship to corporate histories or government histories (databases, emails, memorandums, and so forth), less so social histories that would come out of document drafts or still-largely-unimaginable online popular conversations. All of that would rapidly change by the mid-1990s, as users flooded online and created reams of user-generated content: initially hosted on their own web servers, but within a year or so increasingly turning to third-party platforms that they largely had little control over. Digital preservation moved from being the domain of records and archival management toward the more dynamic world of popular publishing. To see how, a brief overview of the web's history shows how quickly it all changed.

The Early History and Growth of the Internet and Web

The internet's origins lie in part with the network established by the American Advanced Research Projects Agency. It is a far more complicated story, but for the one-paragraph history this can suffice.[58] In 1969, the first message was sent across the ARPANET from the University of California at Los Angeles to the Stanford Research Institute. This heralded the birth of the modern internet. Five years later, an open packet switching standard, the Transmission Control Protocol/Internet Protocol (TCP/IP), was introduced. TCP/IP outlined a way that networks could communicate with other networks. By the mid-1980s, TCP/IP was incorporated into the Unix operating system and made free and accessible to all. The ARPANET also adopted this broad protocol and would eventually be subsumed into the much broader emerging network. The world had converged on a common standard. As more and more servers and computers joined the internet, new means of connecting users to information took shape.

The internet can be used in diverse ways. Today, we primarily interact with the internet using web browsers, using the HyperText Transfer Protocol, or HTTP, to visit websites such as Twitter.com or Facebook.com. Indeed, so much of the internet is delivered by the web that "internet" and "web" are often considered synonymous. Internet users also transfer emails using the Simple Mail Transfer Protocol (SMTP), exchange files using the File Transfer Protocol (FTP), or access remote servers across the network using Secure Shell (SSH), or newsgroups via Usenet. Throughout the period discussed in this book, there were other protocols that used the internet and were thus web "competitors" in the race to get people connected to it. These included the Wide Area Information Server (WAIS) and the Gopher protocol that helped users find information in databases and servers across the internet. Many of the acronyms above, such as HTTP, FTP, SSH, SMTP, will be unfamiliar to most internet users today—testament to the seamless power and easy-to-use interface of the web.

The web quickly rose to be the dominant method for engaging with internet-hosted content. It happened quickly. In 1989, while a fellow at the European Organization for Nuclear Research (CERN) in Geneva, Switzerland, Tim Berners-Lee implemented a system that he had been exploring in earlier forms since 1980. This was a hypertext-based method for navigating networked information that in 1990 became initially known as the WorldWideWeb. On 6 August 1991, web browsers and code for running a web server were placed on the internet and publicized. From that point onwards, anybody could create their own web server. Uptake, however, was relatively limited until 1995. That did not mean, however, that people were not online in other ways.

For the web was not the full story of early, user-driven networked communication. As Kevin Driscoll vividly illustrates in his book *The Modem World*, between the late 1970s and roughly 1995, the Bulletin Board System (BBS) was the "prevailing form of social computing for PC users."[59] This archipelago of individual BBS, each run as a private fiefdom by a local system operator, became increasingly interconnected throughout the 1980s. Users had access to various multiplayer games, arrays of shared text files, and messaging, which helped to pave the way for the subsequent mass adoption of the web. For everyday users, it was the local BBS that would represent the promise of networked communication, as opposed to the distant ARPANET, far removed from a user's personal computer.[60] A vibrant community took shape across BBSes. Rather than simply being supplanted or replaced by the web, by 1995, when the web became increasingly dominant and accessible, many BBSes either transformed into Internet Service Providers (ISPs) or moved their communities online ("a continuation of the modem world," per Driscoll).[61] I will return to BBSes shortly as their ephemerality is a cautionary note for digital preservation.

The web quickly grew. The future of the web was made clear on 30 April 1993, when CERN declared that there would be no fees or royalties imposed on the web. In 1993, too, the NCSA Mosaic Browser—similar to modern-day web browsers—was released at the National Center for Supercomputing Applications (hence NCSA), leading to dramatic improvements in the web's usability and spurring corresponding exponential growth in the number of people hopping on to the now-hyped "information superhighway." By 1996, some 22% of Americans had used the web, up from 14% the year before. More importantly, the web's popularity led to reduced usage of Gopher, WAIS, and other internet protocols. In 1995, only 21% of internet users had used the web; by 1996 it was 73%.[62] This is worth underscoring, as throughout the period discussed in this book, the web may have existed but was not necessarily the dominant method of internet access.

Despite being called the World Wide Web, it is worth making the necessary proviso that the web was certainly not "World Wide" in this period. As Gerard Goggin and Mark McLelland noted in their 2009 volume *Internationalizing Internet Studies: Beyond Anglophone Paradigms*, the "preoccupation with theorizing the Internet as a singular cyber 'space,' as 'virtual,' or as 'deterritorialized' and 'borderless'" has hindered the broader understanding of "diverse international Internets."[63] Similarly, their expansive 2017 volume on global internet histories, bringing together thirty-six chapters of case studies and theoretical interventions, made the strong case that the emerging field of internet histories needed to be constituted as a global one. A linguistic and global shift took place on the internet throughout the 1980s and 1990s. If most of the material was English until the early 1990s, that shifted in the "late 1990s when its percentage of overall language use was reduced by the rapid upsurge in Japanese-, Spanish-, and Chinese-speaking users."[64] Yet even today the Western origins of the internet and much computing infrastructure can still be seen. As per Goggin and McLelland: "It is still somewhat astonishing that the QWERTY keyboard—originally developed to avoid the jamming of commonly occurring letter combinations in English words . . . in nineteenth-century typewriters—is *still* the main human/computer interface even in countries like Japan where the Roman alphabet is not used for daily communication."[65] This is related to the broader arguments around "ASCII imperialism," whereby countries and cultures around the world were forced to use a standard set of characters that ignored diacritics and vowel sounds primarily used in non-Western languages.[66] Or, in other words, as scholars Daniel Pargman and Jacob Palme rhetorically asked as they coined the term, "Why do people using languages other than English so often have problems using the Internet?"[67]

These arguments help us understand the world in which web archiving and the broader field of networked digital preservation arose. It was not a truly global "world" wide web, but rather one primarily rooted in North America and Western Europe. Given the resources required to preserve the web (storage, high-bandwidth network connections, and specialized expertise), it would not be surprising that the earliest web archiving operations would arise in comparatively wealthy, Western countries such as Canada, the United States, Australia, and Sweden.

One last issue is worth exploring: the unique nature of the web itself, largely because of the hypertextual connections that made it a "web." Hypertext was designed to aid memory. However, the web grew so quickly that this itself quickly presented an unprecedented preservation challenge.

A Memory Machine Designed to Forget: The Particular Ironies of Web Preservation

The web's scale has boggled since it exploded in the middle of the 1990s. "Another principle of digital civilization," notes journalist and cultural critic Virginia Heffernan, is that "*No human can read it.*"[68] Or, as the late Grateful Dead lyricist (turned cyber activist) John Perry Barlow noted, reading the internet is like "drinking from a fire hose."[69] That the web forgets is an ironic twist as its hypertextual origins lie in the quest to enhance human memory.

Scholars have long tried to organize the world's information. At its peak in the first few centuries of the first millennium, the Library of Alexandria had "over half a million scrolls," and while myth holds that it burned down, the reality was that its loss owed not to Julius Caesar or the torch but rather to "the work of many generations, Christian and Muslim, who felt no responsibility to care for pagan learning."[70] Human apathy is a powerful destructive force.

Future librarians would carry on this mission of universal access to knowledge. In 1895 Paul Otlet and Henri La Fontaine worked together to "prepare a bibliography of the publications of *all* time and places" (emphasis is mine).[71] In 1910 this led to the creation of the "Mundaneum" in Brussels, dedicated to the collection and organization of all of the world's knowledge, which by 1934 had swelled to 16 million cataloging cards. This Répertoire bibliographique universel, or RBU, was "organized to respond to two basic questions, What works have been written by this or that author? and What has been written on this or that subject?"[72]

The Mundaneum also gave rise to the Universal Decimal Classification, which was a cataloging system that would divide all human knowledge into a set of ten classes, of which each class had ten groups, then ten divisions, then subdivisions, continually allowing for subdivision and thus the expansion of human knowledge.[73] While UDC lives on as a cataloging system, the Mundaneum was forced to move under the Nazi occupation of Belgium, and it faced challenges over its coming years. As a concept, however, it held out the idea that *all* human knowledge could be organized and accessed. It also served as a utopian vision of a centralized internet, anchored in long-term thinking rather than the short-term horizons of corporate stockholders or technology billionaires.[74] Similar themes appeared in H. G. Wells's late 1930s idea of a World Encyclopedia—"World Brain"—a resource that would "be alive and growing and changing continually under revision, extension and replacement from the original thinkers in the world everywhere," connecting universities, researchers, students, and even journalists.[75]

Hypertext was a conceptual attempt to make sense of abundant information itself, relying not on cataloging but rather the essential connection of documents and concepts together. The idea of hypertext stems from Vannevar Bush's 1945 "Memex" concept. Bush was then director of the US Office of Scientific Research and Development during the Second World War and overseer of the Manhattan Project. The name itself was "mem(ory-)ex," or "Memex instead of index."[76] Bush introduced his idea in a famous July 1945 *Atlantic Monthly* article, where he imagined a user connecting ideas and following "trails" between documents, working at a desk, weaving a web of information.[77] The main technology underpinning the Memex was microfilm, which had similarly inspired H. G. Wells's—his vision that "in the near future, we shall have microscopic libraries of records, in which a photograph of every important book and document will be stowed away and made easily available"—stemmed from the same moment of optimism that suffused Bush's work.[78] The Memex was a proposed solution to information abundance. With a Memex, the user could move between information along information webs, making connections, and in so doing would theoretically mimic the mind's information-seeking behavior. The Memex was never built, but the concept endured.

In 1965, Bush looked back at the Memex in "Memex Revisited."[79] Two decades after his original publication, Bush continued to believe that a revolution was necessary in information production, storage, and accessibility.[80] In the future ("a long time from now") Bush foresaw the development of a personal Memex as something that could truly unleash human potential through the augmentation of human memory. Bush was blunt about what this would mean: "Adequately equipped with machines which leave him free to use his primary attribute as a human being—the ability to think creatively and wisely, unencumbered by unworthy tasks—man can face an increasingly complex existence with hope, even with confidence . . . He has built a civilization so complex that he needs to mechanize his records more fully if he is to push his experiment in its proper paths and not become bogged down when partway home by having overtaxed his limited memory."[81] This advanced an argument that struck to the core of historical understanding. Bush saw discontinuities in the historical record as primarily caused by "historical accident" of human memory fallibility; something that could be solved by machines. It was a vision of machines as perfecting human memory-keeping and by extension our historical understandings. As Wendy Hui Kyong Chun has argued, the Memex conflated "storage with access, of memory with storage, of word with action . . . This belief also depends on our machines as being more stable and permanent, and thus better record holders, than human memory."[82]

Around the same time that Bush was reflecting on his Memex, Theodor Holm (Ted) Nelson built on Bush's concept and coined the term "hypertext" itself. In 1965, Nelson defined "hypertext" to refer to "a body of written or pictorial material interconnected in such a complex way that it could not conveniently be presented or represented on paper."[83] Nelson's idea would take shape in his largely conceptual Project Xanadu.[84] As a concept, Project Xanadu offers a compelling hypertext alternate vision that demonstrates that the web's ephemeral nature was not inevitable, at least not in theory. Xanadu's hyperlinks would be "bivisible and bifollowable—capable of being seen and followed from the destination document as well as the originating document," as opposed to the "one-way hypertext" of the web.[85] In other words, these links would never break. There would be no 404 errors. Documents themselves would be continually updated and versioned (akin to Wikipedia), and links would themselves be attached to specific characters. Rather than linking to a URL, users would link to the information or concept itself, stored in multiple places around the network.[86] While Xanadu was not then built, it outlined a conceptual approach that would have seen the long-term stewardship of information.

Another web alternative, the Wide Area Information Server (or WAIS), designed in part by Internet Archive founder Brewster Kahle and made available in 1990, presented a different vision to access the internet. While I will explore WAIS in detail in chapter 3, Kahle's recollection is worth quoting at length as it sheds light on WAIS and the later Internet Archive:

> The web was just so broken. I mean, it's such a simple system, but it boy, it really didn't have any of the features that I put into WAIS towards making it work as a library, and [laughs] with having things live in multiple places and, you know, all the redundancy and structures that I put in place in WAIS, were not in the web. Now, the web works, and you know, WAIS went by the wayside. So, you know, don't be too complicated here. But I had to fix it by going and trying to gather it all back up again, which was really kind of a kluge [an ill-fitting assemblage of parts] and . . . so anyway, that was the first service of the Internet Archive.[87]

Many of these alternate systems had archives, versioning, and other mechanisms that kept data accessible and robust. That did not necessarily make them better. These memory mechanisms contributed to the systems' complexity, perhaps mitigating their growth. Yet it underscores the web's original sin: it forgets too easily. These alternate visions let us understand link rot not as an inevitability but rather the product of decisions. Such decisions may have been logical, or even necessary to achieve scaling, but the web by its nature did not necessarily need to forget.

If the web was used mostly by corporations or governments with robust document retention strategies, this would not have been a tragedy. In these cases, ephemeral documents could have been managed by document managers. However, by the mid- to late 1990s, much of Western and later global society moved online. These were initially affluent college students, then the upper middle class, but then increasingly more diverse societal strata. The web quickly became part of human society and culture, and by extension, our social and cultural record. A broken link would soon be less a momentary inconvenience but rather a symptom of our society's imperiled collective memory. Storm clouds of the digital dark age were looming.

The Memory Hole of Pre-Web Networked Culture

The digital dark age did not start with the web. Pre-web communications technologies had already encountered, and then been lost in, a virtual memory hole. In the introduction to her edited volume *Social Media Archeology and Poetics*, Judy Malloy outlines dozens of historic social media platforms: from 1970s communities in the early ARPANET Request for Comments process and early networked mailing lists and electronic bulletin boards to 1980s Bulletin Board Systems (BBSes) and Usenet discussions (both platforms that started in the late 1970s but hit their stride in the following decade).[88] With exceptions—Usenet is to some degree accessible through a variety of online portals and downloads that have been (in some cases intermittently) available through the Internet Archive, and BBS files can be found at textfiles.com—much of this information is accessible only through oral histories, contemporaneous descriptions in computer magazines and books, and perhaps the occasional serendipitous upload on a hobbyist site.

The vibrant "modem world" of BBSes presented perhaps the perfect storm when it came to being archivally resistant. BBSes are studied through scant traces: "shoeboxes of floppy disks, stacks of waterlogged magazines, deaccessioned library books, unmarked VHS tapes, and pirated software."[89] This is thanks to the ephemerality of a BBS, as Driscoll explains: "Bulletin board systems were never crawled by Google or captured by the Internet Archive's Wayback Machine . . . When a BBS went offline, it vanished from cyberspace, taking all of the files and messages created by users along with it. Data storage was expensive. Most sysops either could not afford or did not think it was necessary to create backup copies of their systems. Often, the best way to record the visual culture of a BBS involved pointing a camera at the computer and making a literal 'screen shot.' Home videos occasionally turn up on YouTube."[90] For all the web's ephemerality, its open standards and ethos would make it possible for a website to be archived by a third party. The later robots.txt protocol, which would give a web administrator a "veto" power over being included in the Internet Archive, gave a bit of the power of the old school sysop power to a webmaster. Active withdrawal was a conscious decision, however, and the default state was not to be opted out. The black hole of the BBS experience would later lend credence and energy to the web archiving experience, as it showed just how precious and vulnerable a widespread virtual community could be to benign neglect and deletion.

Textfiles.com, founded by online archivist pioneer Jason Scott in 1998, contains almost 60,000 files as a "history of writers and artists bound by the 128 characters that the American Standard Code for Information Interchange (ASCII) allowed them," providing an unparalleled understanding of (mainly) BBS subcultures as well as other forms of early networked communication. Focusing primarily on files from the 1980s, Scott notes that "even these files are sometime [*sic*] retooled 1960s and 1970s works, and offshoots of this culture exist to this day."[91] "Textfiles" as a genre were a specific subculture of electronic literature; Driscoll describes them as "the broadsides of the modem world, an ephemeral street literature for the information superhighway."[92] Textfiles.com's invaluable early collection was made possible because of the collecting behavior of Scott himself, as he recalled in a 2012 keynote talk:

> I've always really been interested in computer history . . . Because of a divorce that occurred early in my life, I really got a sense that nothing was permanent. That anything could change . . . And so, when I was very young, when I could call computer bulletin board systems or go onto online services, I would download everything I found and I would put them on floppy disks and have thousands and thousands of text files and programs and images from that period of time. And in 1998 I put up textfiles.com where I had taken these tens of thousands of files and put them up.[93]

The files are interesting reading, from fan-circulated Monty Python lyrics to political debates to conspiracy theories and drug-related discussions. Yet these files are largely divorced from their original context. Even a BBS simulator (such as "telehack") does not capture the interaction and conversations of the early age of networked communication. For that, one must turn to the "history" section of the site, dozens of uploaded narratives reminiscing on their experiences.

This is not dissimilar to other platforms. Consider the PLATO network. PLATO began in 1960 and by the 1970s had expanded to thousands of terminals around the world, networked together via large mainframe computers. Recent scholarship has vividly demonstrated the significance of PLATO (and similar systems) to the origins of online communities. These were platforms where several computing milestones were reached *decades* before the popularization of networked communication via the web. Joy Lisi Rankin has termed the users of these networks "computing citizens," who not only used these platforms, but built them, wrote programs for them, and built communities to engage with each other.[94] It is a story not of Silicon Valley but rather of K–12 schools, universities, and the liberal arts where these networks were founded. Yet, as Rankin has noted in her book, there was also additional value to studying these educational settings; that particular audience meant that publications "often included meticulous details of individuals' encounter with the terminal, the keyboard, the screen, the language, the lessons, the appropriate syntax, and similar issues."[95] Yet it has been a hard history to study in part because of the difficulty of studying now-defunct networks. As Brian Dear explains, "it's as if an advanced civilization had once thrived on earth, dwelled among us, built a wondrous technology, but then disappeared as quietly as they had arrived, leaving behind scraps of legend and artifacts that only few noticed."[96] To write his history of PLATO, Dear interviewed hundreds of participants and corresponded with thousands more. Rankin drew on diverse textual sources as well as a series of oral interviews to bring her story together. All of this underscores the need for robust digital preservation to help document the rich histories of networked conversation.

Conclusion

Human memory is fragile. Digital technology offered up the potential of an information apocalypse, as partially realized in the scant traces left behind by previously vibrant systems and platforms such as PLATO and BBSes. Few primary sources of those communities remain. Yet the digital also opened possibilities: preserve it all and we could have fuller, better human memory. The web in some ways offered both models. It was, by virtue of its decentralization, accidentally fragile and ephemeral. Yet it could be archived in a way that previous systems could not, thanks to its openness and consistent standards. If a web browser could access a server, perhaps a way could be found to remotely back it up.

One final proviso is in order. Despite being called "web archives," from the perspective of a library or archive they might be formally understood as "web collections." Websites are often understood (legally, if not ethically) as akin to publications. As a result, most web archiving has been carried out institutionally by libraries rather than archives.[97] Yet even if the operational role of web archiving largely ended up in libraries, many of the early theoretical discussions occurred through the prism of "archiving." This was not just the normal abuse of the word "archive" but rather a consideration of the institutional role of memory preservation. In this sense, the word was used expansively along the lines of "what should our society preserve for future generations?" Users thought of web archives as "archives" as they filled a similar role. Indeed, during the earliest periods discussed in this book, archivists and record managers were more central to the conversations than they are today. Web archives are thus best understood as operating at the intersections of library and archival science. This adoption of archival techniques by nonarchivists is part of a long tradition of archival work being "taken over by people who were not called archivists," as Markus Friedrich has put it.[98]

A new balancing point needs to be found in the digital age. Archives and libraries are selective not simply because of limited shelf space and attention but also as they try to create usable collections and records that shape historical knowledge. Abundance is not a virtue in and of itself. Too much information obscures rather than illuminates. Add to this the complexity of people who may want to be forgotten, and we can see the shape of a very complicated landscape that came into relief in the mid- to late 1990s.

As the web began to gain in popularity throughout 1994 and early 1995, this story could have gone either way. We could have faced the mass disappearance of information, or perhaps the continuance of the digital preservation field to ensure its preservation. For it was increasingly clear that a digital dark age was looming. But unlike earlier platforms, this one could in theory be saved. Would society rise to the challenge? In the next chapter, I look at how the digital dark age was articulated—and how solutions took shape.

CHAPTER TWO

From Dark Age to Golden Age?

The Digital Preservation Moment

By the mid-1990s, it was clear to many librarians, archivists, and technologists that the web was both fragile and ephemeral. But what to do about it? The early conversation on digital records, initially centered on corporations and institutions, took on increasing urgency when the records of everyday people were considered and the broad impact of a digital dark age was apparent. While we have seen how conversations around electronic records surfaced by the 1960s and 1970s, by the 1990s what had been a debate and discussion largely happening between record managers and within the archival profession emerged into public consciousness through the frame of a digital dark age. This concept made concerns around digital obsolescence seem like a problem that was not just for the Fortune 500 and governments but all of society. Ideas spread outwards from academic venues and fora such as research libraries and the 1994–1996 Task Force on Archiving of Digital Information to have broader social and cultural impact.

Between 1995 and 1998, a series of individuals—including science fiction author Bruce Sterling, Microsoft Chief Technology Officer Nathan Myhrvold, information scholar Margaret Hedstrom, technologist Brewster Kahle, documentarian Terry Sanders, and Long Now Foundation founder Stewart Brand—reshaped the cultural conversation to broaden digital preservation from an academic field to one understood as having wide-ranging implications. This would not just be the preservation of technical or corporate documents but rather the collective digital memory of our society. They thus laid the foundations to avert the digital dark age. Many of the academic discussions explored at length in chapter 1 received their popularization through these people. This chapter explores the process that built a social and cultural consensus about web archiving. The specific history of the Internet Archive, which owes itself in some ways to the unique milieu of Brewster Kahle, will be primarily explored in chapter 3.

This chapter is primarily focused on thought leaders rather than everyday people. While in part this reflects the record—the individuals discussed in this chapter left behind a rich documentary record in film, media, and academic proceedings—it also reflects the reality of digital preservation at this time. Regular users would have seen digital obsolescence as perhaps inevitable, or something that would be tackled on an individual level: to print off a document or save information on a new computer. Even as late as 1996, users still needed newspaper and magazine explainers to trace out the meaning of the 404 Not Found error. One article tried to make the case that the 404 error had entered the popular vernacular—"he's 404, man" for "someone who's clueless."[1] Overall, the ephemerality of web resources was something that was still being explained to many users at this time.[2]

There were other approaches not discussed at length in this chapter. Fan and community archives began to appear on the web in the late 1990s, such as Jason Scott's aforementioned textfiles.com in 1998 or the new media arts organization Rhizome's ArtBase in 1999 (discussed later in this chapter). These vividly demonstrated how some organizations and institutions were independently understanding the web as an archival medium in and of itself, absent state leadership or in many cases respect for copyright.[3] But, ultimately, the coordinated, long-term perspective of the Internet Archive and other projects distinguishes the discussions in these chapters. It was people like Myhrvold and Brand who realized that the preservation of web-based digital material required a "systems approach." Yet everyday users would be profoundly affected by these conversations. Their content would eventually be crawled by the Internet Archive and national libraries. They would feel the impact of the conversations explored here.

If in 1990, the term "digital preservation" did not formally exist and conversations were happening in niche technical venues, by 1997 the idea was so commonplace that a prominent documentary like Terry Sanders's *Into the Future* could be made about the topic. The need for, and existence of, the field was assumed knowledge by information professionals, and numerous conferences had by then been held on the subject by a wide variety of stakeholders. Much of this now drew upon a utopian faith that not only would the historical record eventually be saved but that it would be a dramatically better one thanks to technology. The digital dark age would be averted. Through technology, the mists that occluded the historical record could perhaps be dispersed, leading to better histories. As part of this, the conversation dramatically expanded to include technologists, artists, and writers, who began to see *potential* alongside fear. In averting a digital dark age, could we instead rather see a golden age of memory?

Meeting the Challenge of Digital Preservation: The Challenge of Networked Information

As memory institutions, research libraries and archives—both research and national institutions in affluent countries—had by the 1990s begun to build capacity to meet the challenge of digital preservation. Their earlier professional engagement with electronic records would be helpful. This was fortuitous as they would soon face the difficult problem of preserving the web.

The 1990s would mark a watershed moment for the field with the newly coined name of "digital preservation."[4] Terry Cook, an archivist and theorist then based at the National Archives of Canada, posited in early 1992 that archives were then on the cusp of the second generation of electronic records. As Cook explains, the first generation dated from the 1970s and 1980s, involving digital objects such as surveys, statistics, and censuses. While we saw their complexity in the previous chapter, these types of first-generation records were comparatively simple as they were "flat files" without many dependencies. "Each flat file, with sufficient documentation," Cook noted, "could readily be reconstructed to 'run.' "[5]

The second generation of records, appearing by the 1990s, would be even more difficult to preserve. These new files, as Cook put it, were "large hierarchical, networked, and especially relational databases [where] information is stored in many internal tables, entities or structures, that have meaning only inasmuch as they are related to each other."[6] The sources of information were also shifting from tabular data to "letters, memoranda, policy summaries, operational case files, crucial financial spreadsheets, vital interpretive graphic material, even maps, photographs and sound recordings . . . being converted into the digital bits that make up electronic records."[7] The groundwork for these complex records was laid well before the web, but these early experiences would help set the stage for the web's subsequent preservation.

One of the first reflections on the unique digital preservation challenge ahead came with "Electronic Technologies and Preservation," a report written by Yale University Library administrator Donald J. Waters. Presented at the 1992 Research Libraries Group annual meeting, the paper articulated a future shape of libraries and archives in the digital age. The growing amount of digital information presented growing access problems. "Information also is increasingly available electronically as a direct source of recorded knowledge," Waters argued, noting the added challenge of "compound documents," which included hypertext, "mixed text and image," and multimedia.[8] Complementing Waters, in September 1992 the final report of the Cornell/Xerox joint study on digital preservation was released. It foresaw digital technology as enhancing access, highlighting a demonstration project that showed how one could remotely access digital images over a network, echoing earlier utopian takes on networked information.[9]

As increasing attention was paid toward the merits of making information available on the internet and web, it was not long before digital preservationists began to worry about the web's long-term sustainability. Yet there was still a gap of a few years between the early 1990s buzz around networked resources and the rise of fears around its ephemerality five or so years later. In part, this reflects the time needed for the web's dominance to be clear, as it was still part of a broader ecosystem of internet access platforms including Gopher and WAIS.

Nineteen ninety-four was also the year the Commission on Preservation and Access teamed up with the Research Libraries Group to launch the Task Force on Archiving of Digital Information. The Task Force would produce a series of reports throughout 1995 and 1996, culminating in a well-received final report. The twenty-one-member Task Force included librarians, private-sector representatives, and publishers who collectively explored key problems in the field intending to ensure "continuing access to electronic digital records indefinitely into the future."[10] This broad membership was key. As Task Force member Hedstrom recalled "it tied research libraries and archives and digital preservation in a way that was quite holistic for its time."[11]

The Task Force's final report explored themes that would influence the digital preservation world. The report opened with an evocative image, presaging the rhetoric of a digital dark age: "Today we can only imagine the content of and audience reaction to the lost plays of Aeschylus. We do not know how Mozart sounded when performing his own music. We can have no direct experience of David Garrick on stage. Nor can we fully appreciate the power of Patrick Henry's oratory. Will future generations be able to encounter a Mikhail Baryshnikov ballet, a Barbara Jordan speech, a Walter Cronkite newscast, or an Ella Fitzgerald scat on an Ellington tune?"[12] The digital age was dawning, the report noted, with "virtually all printing and a rapidly increasing amount of writing" being done digitally. Yet this needed to be considered in a context of ever-increasing complexity. Digital information's life might become, echoing Hobbes, "nasty, brutish and short."[13] While the Task Force recognized the long history of digital preservation, the web was an accelerant. Everything would soon be transformed.

How could a library or archive preserve a hyperlinked networked resource if one could not preserve all of the other resources that were hyperlinked from it? Was this content not an integral part of the document? Indeed, the Task Force noted the challenges brought on by an interconnected resource like the web. The report used a network metaphor to explore the question: "If the integrity of these objects is seen as residing in the network of linkages among them, rather than in the individual objects, or nodes, on the network, then the archival challenge would be to preserve both the objects and the linkages, a task that would today be exceedingly complex." The "stop-gap measure would be to treat the network in terms of its component parts and to take periodic snapshots of the individual [web] objects."[14] This was an early preview of the defining challenge of web preservation: hyperlinks and the issue of completeness.

By 1997, then, thanks to these efforts, the importance of preservation was now understood by many in the library and archival field to be a defining challenge. Two highly cited and influential papers illustrate this. One of them was by an individual long involved in the electronic records field, bringing earlier expertise to bear on this new challenge. This was Hedstrom, by then an associate professor at the University of Michigan's School of Information and a Task Force member, who minced few words in a 1997 article entitled "Digital Preservation: A Time Bomb for Digital Libraries." Hedstrom noted that "new technologies for mass storage of digital information abound, yet the technologies and methods for long-term preservation of the vast and growing store of digital information lag far behind."[15] The intersection of "mass storage" and "long-term preservation" was, to Hedstrom, a ticking time bomb. As Hedstrom recalled, this was her most-cited publication, which, to her, was surprising given it was in the somewhat niche journal *Computers and the Humanities*: "That piece, I would say, was kind of putting a stake in the ground based on where things were at the time. And I have to say, in all honesty, for me personally, it was kind of like, OK, this is the problem and, you know, I don't have a lot more to say . . . I don't have a lot more to say about this."[16] The time bomb was a provocative metaphor. Act now, or irreparable damage would be done.

The second pivotal paper was "A Digital Dark Ages?" by Terry Kuny. Kuny raised the special challenge of web preservation. "Libraries which seek out materials on the Internet will quickly discover the complexity of maintaining the integrity of links and dealing with dynamic documents," he noted.[17] Kuny recalled this piece as a similar sort of summative, state-of-the-field article: "[It was me] speaking to the converted in some respects, but saying, 'basically, this is a huge challenge. I don't know that we're up to it.' [laughs] I don't know that anybody is up to it, actually."[18] Both accounts were prescient about what lay ahead and helped set the agenda. A digital dark age was on the horizon, and society needed to act quickly.

Within a few years, research libraries would understand digital preservation as a core task. As the web rapidly grew, the conflation of intellectual activity and the web's growth would lay the foundation for libraries to move into action. Would they be ready in time? For there were now growing fears around the idea of an irrecoverable cultural loss. Concerns mounted that as people moved onto the web, their information could quickly disappear. Would the web be a place where knowledge went to die?

The Dead Media Project

Lisa Gitelman notes that "all media were once new."[19] Just as we continue to call the web "new media," the phonograph was just as cutting edge and new in its heyday. New media becomes old media, which in turn can become—as science fiction author Bruce Sterling evocatively put it in 1995—*dead* media. Sterling presented his "Dead Media Project" as a way to challenge the "newness" of new media. Sterling and his message would introduce ideas of digital preservation and obsolescence to a much broader audience than those reached in scholarly journals and professional task forces. The way that Sterling articulated the problem, digital preservation was not just an intellectual or academic concern. It was a problem for society.

Addressing the Sixth International Symposium on Electronic Art in Montreal, Sterling coined the phrase "dead media." This was an evocative and influential framing. As Tara Brabazon argues, the "term captured lost, marginalized, and obsolete media. It was part archive, part nostalgia, part requiem."[20] Sterling's speech, subsequently published as the "Life and Death of Media" manifesto, articulated the ephemerality of digital media. "Before we install the latest hot-off-the-disk-drive version of Windows for Civilization 2.0," Sterling argued, "we ought to look around ourselves very seriously. Probably, before leaping into postmodern ecstasy into the black hole of virtuality, we ought to make and store some back-ups of the system first."[21] To do this would require a rethinking of society's relationship with technology.

Sterling argued that society needed to move past an implicitly Whiggish narrative of "technological history." In other words, the model of ever-increasing improvement toward an enlightened present inevitably left little room for technologies that were not part of the main narrative. Sterling contested the techno-utopian narrative of unfettered progress. He argued that media history was governed by a paradigm of progress: "all technological developments have marched in progressive lockstep, from height to height, to produce the current exalted media landscape."[22] What if somebody wrote a history of all the inventions that did not fit into this narrative of progress, to instead consider the new media that became dead media? Sterling proposed that somebody (not him: "someone else") should write *The Dead Media Handbook*, "a field guide for the communications paleontologist."[23]

Sterling made the fragility of digital information clear to his artistic audience. At one point in his speech, Sterling gestured at his laptop computer—"a Macintosh PowerBook 180." He noted that it was an impressive machine but that ironically the "name PowerBook somehow suggests that this device can last as long as a book, though even the cheapest paperback will outlive this machine quite easily."[24] This was important. Sterling continued:

> Suppose you compose an electronic artwork for an operating system that subsequently dies. It doesn't matter how much creative effort you invested in that program. It does not matter how cleverly you wrote the code. The number of man-hours invested is of no relevance. Your artistic theories and your sense of conviction are profoundly beside the point. If you chose to include a political message, that message will never again reach a human ear. Your chance to influence the artists who come after you is reduced drastically, almost to nil. You are inside a dead operating system.[25]

In other words, to Sterling, you "have become dead media." Something needed to be done.

The energy around his call to action led to the formation of the Dead Media Project. The eloquent and forceful nature of Sterling's speech gave the dead media manifesto enduring life beyond Montreal. The following year, in 1996, Sterling, joined by fellow science fiction author Richard Kadrey, cofounded the Project. In their coauthored "Modest Proposal and a Public Appeal," Sterling and Kadrey made the case that new media *does* die. As they explained, everybody knew of newspapers, TV, video, cable, but perhaps not the "Edison wax cylinder," "The Pandorama," or the "teleharmonium." Just as businesses no longer used pneumatic tubes to send information, Sterling mused, "How long will it be before the much-touted World Wide Web interface is itself a dead medium? And what will become of all those billions of thoughts, words, images and expressions poured onto the Internet? Won't they vanish just like the vile lacquered smoke from a burning pile of junked Victrolas?"[26] Through these statements, which articulated what was at stake, Brabazon argued that Sterling "granted the internet a history and ensured that it was part of a wider analysis of media, communication and identity."[27]

The intellectual ferment and energy behind the project's conception would ultimately be more important than its execution. The project's listserv grew to have around 600 active individuals by 1999, a place where members could submit examples of dead media, which Sterling would subsequently edit and distribute.[28] Yet, in an ironic twist, the project itself languished and began to degrade through neglect. While it has been partially restored today, Brabazon's note that "if there is anything sadder than dead media, then it is dead links from a Web site on dead media" rings true.[29] The community had fallen victim to digital obsolescence.

Yet the project provided crucial historical context and awareness of the web's ephemerality. That the Dead Media Project itself became obsolete does not occlude its intellectual contribution. Sterling made it clear that, given the ephemerality of so much new media, he did not "expect the Web to last very long indeed, at least not in its present form." After all, even by 1999 as he explained in an interview with a new media journal, "there are large numbers of abandoned websites on the Web that were partially constructed and then left to rot in cyberspace. And have you tried using 'gopher' or 'WAIS' lately?"[30] Under Sterling's model, preservation was articulated as less a default outcome and more an exception. It was another blow for the Whiggish vision of progress. What would this, however, mean more broadly for our historical record? Would we be on the verge of a digital dark age? Sterling suggested that new platforms—before they could reach a critical mass and become ubiquitous—were especially risky: it was still unclear what direction the web itself would take. What if everybody built a vibrant culture on the web, only to lose it all? Would we be witness to a mass erasure of history?

The Specter of a Digital Dark Age

Enter the idea of a digital dark age. The concept was best articulated in a January 1995 *Scientific American* article by RAND Corporation researcher Jeff Rothenberg. Rothenberg explored the difficulties in preserving digital data. What if, in fifty years, his grandchildren found a CD-ROM? Could they read the physical medium, and even if they could, what about the file formats within? This was not a new challenge. Rothenberg nodded toward the apocryphal fears of the 1960 American census. But as everyday people began to move to the "digital," he argued that this was going to become an increasingly pressing issue.[31] Writing for a popular audience, Rothenberg evoked a mental image that would dominate the popular understanding of digital preservation for the coming decades.

Echoing Rothenberg in his 1997 presentation to the International Federation of Library Associations (IFLA) conference, Terry Kuny stressed that "being digital means being ephemeral . . . it will likely fall to librarians and archivists, the monastic orders of the future, to ensure that something of the heady days of our 'digital revolution' remains for future generations."[32] Stewart Brand, the technologist behind the late 1960s *Whole Earth Catalog* and later cofounder of the Long Now Foundation, was also raising concerns by the late 1990s that records were being quickly lost. "We can read the technical correspondence from Galileo," Brand argued in 1998, "but we have no way of finding the technical correspondence [of the digital era]."[33]

The use of a historical argument made the digital dark age framing so effective. It was often informed by a personal experience of the digital historical record slipping away. Reflecting almost twenty-five years after his "digital dark age" paper was presented at the IFLA conference, Kuny noted to me that his ideas came out of personal experience: "I realized that, as I started getting a little bit older and moving into the library community, even my own personal digital footprint was disappearing. And it wasn't even accessible to me. I was having my own personal 'digital dark age' all the way through, and it continues. I see it happening all the time. I've had a big digital life, but I haven't been able to maintain the record of my own digital life."[34] Early technology adopters, who would see their own private and professional records face obsolescence before the wide adoption of personal computing, would be key to helping motivate early concerns. The limited spread of personal computing meant that while there was not a popular groundswell of stories about digital loss, early adopters set the stage.

Those working in the archives and records field had, of course, worried about the digital dark age before it became a media trope. Asked about the term, Edward Higgs, a historian who worked at the United Kingdom's Public Record Office in the early 1990s, and who would help drive early scholarly work in this field, joked, "I think I invented the term actually at some point! I mean, it's such an obvious thing to say. So probably hundreds of people were using it . . . We were very concerned."[35] Hedstrom recalled that the term itself "came out of kind of the records management world," reinforcing Higgs's tongue-in-cheek origin story.[36] It was clear that this idea was increasingly widespread by the early 1990s. Others, however, questioned the "sense of panic" inherent in framing digital preservation as a dark age. Paul Koerbin, who began working on the Australian web archive in 1996, recalled that it was not "quite as dramatic as . . . a 'digital dark age.' It was just all this material, [these] publishing formats that we're not collecting." Perhaps naïvely, as he recalled, Koerbin figured that "you know, once we get [these projects] up and running, we can deal with this."[37] Koerbin recalls that it was viewed as an opportunity: "I heard a lot more about visions of being able to 'time travel' through the past web rather than falling into a black hole," portending the prospect of a golden age of memory.[38]

We have seen these kinds of source gaps before, even if they lacked the evocative framing of a dark age. For example, television archives continue to be mostly inaccessible for historians. Television broadcasts are "even more ephemeral than the Internet."[39] Not subject to legal deposit, and mostly produced under a copyright regime that required a station to keep just one copy, television has remained mostly off limits and historians' understandings of the postwar world have suffered as a result. But, perhaps because of the power of digital storage and the democratic prospects of digital media, it was digital media—and the web in particular—that ignited fears of a digital dark age. Television was a broadcast medium—the internet and the web were even more complicated because of the publishing dimension. Due in part to the techno-utopian ferment of the time, many commentators assumed that this problem would eventually be solved. Indeed, to most, it seemed like an article of faith. But in the meantime, commentators worried about how long this would take. How long would the "gap" between the adoption of digital media and a long-term preservation solution be?

The potential length of this gap varied. Danny Hillis, who, as we will see, was instrumental in the digital preservation field more generally, noted that "from previous ages we have good raw data written on clay, on stone, on parchment and paper, but from the 1950s to the present recorded information increasingly disappears into a digital gap. Historians will consider this a dark age."[40] As Brand noted as late as 1999, "with digital media it is increasingly possible to store absolutely everything. The traditional role of the librarian and curator—to select what is to be preserved and ruthlessly weed everything else—suddenly is obsolete."[41]

The digital dark age was not just the issue of obsolete disks no longer fitting into a disk drive, or hard drive faults. Those are problems, of course, but solvable ones. As we have seen, the most significant issues are policy and institutions. Hedstrom mused on this point, noting that the "field in general got kind of hung up on some of the wrong things . . . in particular, [it] got hung up on technology, obsolescence and formats."[42] To her, the digital dark age came from the decision for institutions to fix content and keep it offline (and thus inaccessible). Kahle, who would later found the Internet Archive, waxed eloquently at length on this point in conversation with me. The specter of a digital dark age haunted him "every day" of his life, but what he found most interesting was how others understood it:

> It's been interesting to see other people's ideas of what the threat vectors are [of a digital dark age]. You know, how is it going to happen? And it's changed over time . . . It's interesting to see what other people's, you know, what part of the elephant

> of the preservation problem they see. And whether it's the hard drives . . . or is it going to be institutional instability? . . . The biggest problem I see is corporations, the rise of corporations, which is just this viral disease that has really hit the world really since World War Two.[43]

As Kahle explained, "I thought this was a technical problem. This isn't a technical problem." The problem was that corporations don't have long-term perspectives. Copyright further compounded this. Hedstrom recalled this of the Task Force on Archiving of Digital Information: "And so the important point, which I think came from, as I recall, kind of came from Don Waters, was that the intellectual property owners are kind of the first line of defense against losing stuff. Right? And if they can't take care of stuff on their own, then they've got to negotiate somehow on their intellectual property rights."[44] As we will see, the Internet Archive's foundational structure and early activities grew out of both realizations: that for-profit enterprises and copyright lie at the heart of these challenges.

Both Kahle and Hedstrom were correct when they emphasized that the political and economic challenges of digital preservation would be more vexing than technological ones. Indeed, the rate of file format change has slowed. This is thanks in part due to the increased file sharing made possible by the web, which may have led to format consolidation. As Hedstrom noted to me, "the Web was a real boon to just being able to keep things going and accessible and moveable . . . when you now need to be able to exchange things in real time, it becomes much, much easier to exchange things over time."[45] This was hindsight, of course. At the time, it all seemed overwhelmingly challenging. Would our human record go the way of the dinosaurs? At Microsoft headquarters in Seattle, one executive was making that connection.

"Save the Web": Nathan Myhrvold and the Mainstreaming of Web Archiving

A dinosaur brain set Nathan Myhrvold down his path to web preservation. Looking at a "plaster cast of the tiny brain pan of a Tyrannosaurus rex," Myhrvold recalled, "reminded him how few fossil records the dinosaurs had left behind." From there, Myhrvold began to think about the records that we leave behind—and how, in 1996, that would inevitably involve the records humans were posting on the web. "And in a conceptual leap worthy of Mr. Myhrvold's training as a physicist," wrote Denise Caruso in a *New York Times* profile, "this thought set him to worry about the Internet and the World Wide Web."[46]

Myhrvold, Microsoft's chief technical officer between 1996 and 1999, is fascinating for both his sudden arrival and then departure on the web preservation scene. Becoming a leading figure in 1996 amongst web preservationists thanks to a widely circulated memo, Myhrvold gave a publicized plenary address on the topic in 1997, before almost as quickly turning to his many other endeavors and leaving the preservation conversation.

Myhrvold brought an interesting background and perspective to bear on the field's problems. By the age of 24, Myhrvold had already earned a doctorate in math and completed a year of a postdoctoral fellowship with Stephen Hawking at Cambridge University, before pivoting to found a Silicon Valley technology company in 1984. It was subsequently acquired in 1986 by Microsoft. He had then become Microsoft's director of special projects, and eventually chief technology officer. Beyond his everyday supervision of Microsoft's software development portfolio, Myhrvold became "[Bill] Gates's strategic planner and futurist."[47] Internally within Microsoft, he became known as the "Insider as Outsider," releasing "several times a month . . . lengthy memorandums (which can run to nearly a hundred single-spaced pages) that question what Microsoft is or should be doing."[48]

While many of these memos are now inaccessible, victims of email's ephemerality and corporate privacy, those that remain demonstrate an expansive scope. Consider Myhrvold's most famous missive, "Road Kill on the Information Highway," a 20,000-word rumination on the internet's impact. Written in September 1993, the memo grasped the social and cultural impact of networked communication.[49] It captures Microsoft's early grappling with the subject.[50] Crucially, it portended future directions for Myhrvold when it came time for him to consider preserving society's record.

Myhrvold had been reflecting since 1993 on the role that widespread storage of digital information would have on everyday people. The problem of digital preservation was framed in solutionist terms, a solvable problem that in its resolution could go one step further and herald a golden age of memory. "Given the increase in storage on PCs, why not record every version of every file?" Myhrvold rhetorically asked in this memo, "high speed networks and new software will make this quite cheap." Universal storage could be akin to an airliner's black box, applied instead to a wider array of social contexts. Some of this was chilling, as increased storage could, for example, keep archives of surveillance cameras. Myhrvold grappled at length with the downsides (all people have told "a lie or done something that in retrospect they aren't proud of") but noted that "whether putting your life on line is good or bad, it is *very* clear that it will be both feasible and quite cheap. Given this I believe that it will be widely used in at least some circumstances."[51] For a memo written in 1993, it was prescient in how it grappled with the societal implications of storage . . . and the downside of having an ever-present historical record.

Beyond the broad implications of cheap storage, Myhrvold saw computing as central to information distribution. In this, he drew parallels with the printing press that drew on his robust understanding of historiography. "It is estimated that Europe had on the order of ten thousand books just prior to Johan's invention—within fifty years it would have over eight million . . . I believe we are on the brink of a revolution of similar magnitude. This will be driven by two technologies—computing and digital networking."[52] These historical connections were in keeping with his broader historical worldview. Myhrvold was drawing thoughtful comparisons between new information systems, the industrial revolution, and connected historical arguments with contemporary developments.[53]

These currents came together for Myhrvold when he saw the dinosaur brain. The brain spurred thinking about the fossils that our own society would leave behind (it might seem to be a stretch, but as the father of a one-time dinosaur-obsessed child, I can say that reading about dinosaurs does spur thinking of a much vaster time scale!). In March 1996, Myhrvold explained to the *New York Times* why he was so worried about losing web content:

> "The Web is losing its history," Mr. Myhrvold said. But with so much that seems irrelevant published on the Web today, what does it matter? Who needs to chronicle the human achievement of a Web site that is connected by live video feed to a toilet?
>
> Not the point, according to Mr. Myrhvold. Over the last two decades, an historic shift has occurred as an enormous amount of human endeavor—culture, commerce, communication—has moved from the physical world into the realm of electrons.[54]

Myhrvold grasped the web's evolutionary potential for historical research, distinguishing his approach from that of the digital preservation community (which was then somewhat focused on university or corporate records). He considered the forthcoming impact on ordinary people. "Every day the Web becomes more and more important in academics, business and ultimately contemporary culture itself . . . Sure, we were all writing, but if we don't save it, it isn't part of the historical record," Myhrvold explained. Indeed, he wondered if people writing *about* the web in books and magazines might end up being better preserved than the primary documents themselves.[55] As I have seen in my own research into the 1990s web, he was right.[56]

Myhrvold was thus among the first to correctly identify that web preservation was not key to preserving the record of technical decisions and internet culture itself, but *all* culture as reflected in these new digital media. Perhaps because he occupied a front row seat on network debates at Microsoft, he articulated that the web was the new printing press. Drastic action would be needed for its long-term preservation. Our collective historical record was threatened. It was not just preserving internet history but preserving history on the internet. Growing out of this, by mid-1996, Myhrvold was appearing in the media as somebody, as one magazine article put it, who "lately has been championing the idea of an archive" of the web.[57]

This was bolstered by the widely distributed "Save the Web" memo that Myhrvold wrote. In the *New York Times*, John Markoff argued that the "rallying cry to archive the Web began last year when Nathan Myhrvold, the chief technology officer at Microsoft, sent an electronic 'Save the Web!' message to a group of colleagues. 'The Internet isn't naturally archival,' he said. 'The Net isn't going to archive itself.' "[58] Similarly, *Slate*'s Bill Barnes made the same connection, arguing that Myhrvold's "Save the Web memo last year helped start the archive movement."[59]

By spring 1996, Myhrvold's concerns around the disappearing web brought him into conversation with a historian, Philip L. Cantelon, president of the historical consulting firm History Associates Incorporated. Cantelon brought Myhrvold as well as one of the internet's founding figures, Vint Cerf (who, in 1973, had coauthored the foundational TCP/IP protocol that underpins the internet), together to discuss convening a conference to deal with the "danger of losing the documents and information necessary to write the history of our times."[60] They planned to hold the event in late 1997 but "all quickly agreed that the urgency of the problem required more expeditious action and the conference was scheduled for February 1997."[61] Invitations would be sent out, and in February, Myhrvold and Cerf cochaired the "Documenting the Digital Age" conference. This would be a major gathering of people to discuss the next steps for action.

Documenting the Digital Age: Historians and the Turning Point of Web Preservation

The Documenting the Digital Age conference was a significant turning point in web preservation, bringing together historians, technologists, librarians, and archivists from across the private and public sectors. Conference cochairs Cerf and Myhrvold decided that they wanted to move the discussion "beyond the usual professional boundaries," drawing instead on experts from "private and public sectors, specialists in archives, communications, digital technology, history, and the law."[62] A big problem like preserving the web as the future historical record of society would need diverse voices, from the entrepreneur to the archivist to the copyright specialist. Yet there are ironies to studying the conference. Despite its emphasis on ensuring web preservation and developing action plans, the conference materials themselves were not preserved.[63] The conference website, unevenly preserved by the Internet Archive, fell victim to the dreaded 404 after only a few years.

Held between 10 and 12 February 1997 in San Francisco, Documenting the Digital Age was organized by History Associates, with support from the National Science Foundation (NSF), the telecommunications company MCI, and Microsoft. This sponsorship was in itself significant. MCI sponsored it primarily thanks to Vint Cerf, the TCP/IP codeveloper and then senior vice president of technology strategy at MCI. Myhrvold (presumably) brought Microsoft to the table. Donald J. Waters, then the director of the Digital Library Federation, noted that such sponsorship was rare: big business did not typically sponsor such events. As Hedstrom noted to me, this sponsorship was key for the small field. "We were desperate at that time to try to get industry interested in this, and we were talking about things like: 'could we get Microsoft to have, besides having a save button, to also have an archive button?' " Hedstrom recalled, raising the idea of a button that would comprehensively extract a website's source code and send it elsewhere for preservation.[64]

Waters specifically noted Myhrvold's presence, explaining that he "provided its keynote theme in a widely circulated memorandum in which he asked: 'who will save the Net?' "[65] The wide array of attendees was notable, with *Slate*'s Bill Barnes noting that the event gathered "experts from the computing, telecommunication, and archiving worlds to explore these issues."[66] User voices—notably historians—were relatively sparse, a point to which I shortly return.

The conference has gone largely unremarked upon in the literature, perhaps because of its ephemeral digital footprint. The historian Roy Rosenzweig observed that while Documenting the Digital Age was an important event, as a "partial exception" to the trend of historical nonengagement with archives, in that it involved several historians "but only one university-based historian"—the website had "disappeared from the web, [nor] is it available in the Internet Archive."[67] The conference website was not preserved. However, thanks to the relatively recent keyword search functionalities in the Internet Archive's Wayback Machine, I was able to find the postconference website at the Internet Archive. This site provided the basics of talks and structure. Hedstrom also provided me with an extensive array of documents from the conference, ranging from position papers to the NSF final report to correspondence surrounding the event.

Documenting the Digital Age covered many topics, from Myhrvold speaking on the overarching question of "Why Archive the Internet?" to specific points by archivists around what should be preserved, how to preserve it (Brewster Kahle and Donald J. Waters spoke specifically on this question), legal issues, and questions of search. The discussion culminated in the big question: "What do we do next? Who will take responsibility? Who will provide funding?"[68]

Reflecting on the conference, Hedstrom said it was rewarding. She specifically recalled Kahle's energy and enthusiasm, as he had just begun collecting with his Internet Archive. Kahle had also begun to worry about takedown notices and other obstacles being thrown in the way of the Internet Archive:

> Brewster's kind of thing was: "Well, let them come after me. I don't want to be in a position where I have to get prior clearance from anybody who might claim copyright in this stuff and before I capture it" . . . And the discussion was about that kind of thing. Like what would happen if somebody told you that they want their stuff eliminated and wiped out?
>
> And it was more giving Brewster a little bit of: "OK, you have to act a little bit more like a grown up, but you don't have to cave in completely."[69]

For the young field of web archiving, it was optimism—let's save as much as we can—tempered by realism and pragmatics. Such a conversation served as a useful bridge between long-serving practitioners and newly arrived technologists.

The precirculated papers and presentations covered a lot of ground. Kahle stressed that while the "documents on the Internet are easy documents to collect and archive," haste was needed because of the short life span of documents. Otherwise, the web would be too unreliable to cite.[70] Hedstrom imagined a hypothetical researcher exploring Gulf War Syndrome. As part of this thought experiment, the user discovers that there was an online discussion group. Some of the data has been archived only on magnetic tape, requiring a specialized workstation. Ultimately, after a series of obstacles, the imagined researcher abandons the project.[71] Only through a series of forward-thinking interventions could this state of affairs be averted. Waters presented on the experiences of the Task Force on Archiving of Digital Information, echoing the Hobbesian line that the life of digital information would be "nasty, brutish, and short."[72]

Myhrvold's plenary explored how the "Internet is rapidly becoming a key method for communication and the dissemination of documents and ideas," including email, webpages, bulletin boards, chat services, and the rise of indexing services to find all this information. This represented a revolutionary shift in publishing: "All of these aspects of the Internet are remarkably cheap, both in the absolute, and in comparison with other media." Myhrvold's central argument underscored his intervention:

> These properties make the Internet a tremendous information resource. Technological trends suggest that the Internet will get a variety of new capabilities over time, such as the ability to easily deal with high quality video. The Internet is about all you could ask of an information resource.
>
> Except one thing: the Internet is not naturally archival.[73]

There was tension around Myhrvold's utopian vision of trying to save everything. In focusing on conceptual issues, he risked overlooking the very real technical challenges facing the field.

Indeed, much of the final report prepared for the National Science Foundation goes into detail around the extensive debates that came out of Myhrvold's evocative call to "save it all." Hedstrom recalled these debates in our interview. To her, Myhrvold "was sort of standing up talking about how you could save everything. And there's no reason to think about what you save and don't save. And I mean, those of us in the archiving world thought it was pretty, pretty naïve."[74] Myhrvold's argument was: "I believe that it is incredibly dangerous to second guess future generations, and edit the historical record. We should archive all of the net that we possibly can. Ironically, it is probably cheaper and easier to store it all. Digital tape is cheap. Human time to categorize and edit is expensive by comparison. Leave the editing and selection for future generations—or their software agents."[75]

These arguments, as we will see in chapter 3, would be enacted with some success by the Swedish national library. The Internet Archive, too, would eventually try to save it all—while this is an unachievable goal, it was something to aspire to. Debate at the conference on this point was extensive, reflecting the lack of consensus around the right approach to take. To some attendees, mass collecting would be postponing the inevitable process of selection (at the very least, it might need to be done by future researchers). Other participants feared that too much information being collected might obscure the data's context, while others wondered if mass collecting would be the "best use of scarce funds." Conference attendees were aware of the complexity facing them.[76]

This debate suggested a growing divide between a technologist approach to preservation—collect it all and sort it out later—and the traditional professional approach to preservation with an emphasis on curation, context, and descriptive metadata. Myhrvold sketched out a vision of future research processes that in some ways presaged the utopian spirit of the digital humanities a decade or two later. In this hypothetical future, researchers would use technology to navigate information, rather than reading pages one by one or consulting finding aids. Researchers would leverage emerging information retrieval technology. "Want to find out who started a particular idea, rumor or trend?" Myhrvold rhetorically asked, imagining a researcher who could find the first occurrence before moving on to related instances. Or, alternatively, to compare presidential elections in 1996 and 2000 by running "cross comparisons by searching and cataloguing sites . . . [t]raditional historical analysis will be possible, but so will many other new methodologies that are enabled by the information retrieval software."[77] Myhrvold was arguably correct about the falling price of storage but underestimated the challenges of processing and making usable terabytes of raw data. We will return to these debates in the following two chapters, as the Internet Archive and Sweden adopted a "collect it all and sort it out later" strategy, whereas Australia, Canada, and the United States emphasized curated and described collections that could be more immediately useful.

In sum, the conference was a comprehensive event that succinctly explained the state of web archiving in 1997 and considered potential future directions. The major players in the field had gathered for this initial conversation, scoping out the problem for the next generation. The attendee list skewed toward archivists, media (John Markoff from the *New York Times* and Bill Barnes from *Slate*), and libraries. There was one notable gap in the attendee list: few historians.[78] Few historians were there, save MCI's corporate historian (Adam Gruen, a historian of science), James B. Gardner (a consultant with History Associates Incorporated), and an opening keynote address by Rutgers University historian James Muldoon on the communications revolution.

Barnes summarized the event in *Slate*. The corporate sponsors and attendees provided a unique flavor. "Corporate executives complained that because their archives are routinely subpoenaed by plaintiffs' attorneys, they have every incentive to shred their data instead of preserving them," noted Barnes, adding that lawyers also "worried aloud about privacy and copyright concerns." Attendees also discussed the ethical implications behind web archives, beginning a vein of discussion that endures today: "Should you have the right to exclude your public page from the archive? (Consensus opinion: Yes.) Should we be saving usage logs, which detail every page a person sees? (Probably not.) Doesn't this whole thing violate current copyright laws left and right? (Almost certainly.) Should those laws be amended to allow such an archive? (Probably.)"[79] At Documenting the Digital Age, discussions around the importance of preserving the web were percolating amongst a growing body of people. Reading the proceedings today, I was struck by the degree to which these conversations held almost a quarter of a century ago echoed the ones being discussed at today's web archiving conferences.

After the event, there were attempts at organizing follow-up activities to build community and articulate next steps. A shorter (and smaller) follow-up meeting was held in May 1997 in San Francisco, sponsored by the Council on Library and Information Resources (CLIR), with an attempt to create an "agenda for further action."[80] Goals included assigning direct responsibility and continuing these conversations by bringing discussion points back to professional groups, whether governmental, academic, or private. The intention was to keep the momentum going by ensuring that the conversation continued in diverse settings.

Unfortunately, Documenting the Digital Age was not the catalyst the organizers hoped it would be. The lack of the event's long-term impact can be seen in just how difficult it was to learn about the conference despite the high-profile attendees and supporters. Perhaps a combination of Kahle's Internet Archive moving forward, as well as national libraries, meant that the immediate action items were handled? Yet the conversations were important, covering significant topics. The event also helped bridge the generational gap, bringing long-time practitioners into conversation with a new generation. Indeed, a documentary video would soon help crystallize and summarize many of these issues for a broader audience.

Into the Future: The Conversation Goes Broadcast

By 1997, the major currents of the digital preservation problem were increasingly well known and widespread across the library and archives field. Awareness would be bolstered in 1997 with the release of *Into the Future: On the Preservation of Knowledge in the Electronic Age*, an hour-long documentary directed by documentarian Terry Sanders. It aired on PBS and was also sold and circulated on videotape. Drawing on interviews with digital preservationists (Yale's Donald J. Waters, RAND Corporation's Jeff Rothenberg), scholars (MIT's Sherry Turkle and Michigan's Hedstrom), and other web luminaries (Tim Berners-Lee and MIT Media Lab's Michael Hawley), the video aired on PBS in January 1998 following earlier commercial availability in September 1997.[81]

Into the Future is an engaging watch. The documentary opened by touching on the problems of information overload, format obsolescence, electronic books, and literature. It concluded with an in-depth exploration of the problems facing those who sought to preserve the web. The narrator set the stage: "The sheer quantity of digitized information, and the dynamics of an evolving computerized world, create complex problems. One of the most serious is that we pay little attention to preserving electronic writings for the long term, to making sure that important and irreplaceable work will be saved and be available not just for our own use, but for generations to follow. What's increasingly at risk is survival into the future of recorded knowledge, the survival of collective memory, the core of civilization, the human record."[82] The film in part revolves around the hubris of digital creators. As Rothenberg explains in the film, computer scientists tend to "charge ahead into the future" without paying heed to "old, obsolete systems."[83] The film interviewed most of the important people involved and discussed the major issues in the field. Notable missing individuals included Kahle and Myhrvold.

The documentary underscored the particularly vexing preservation problem of the web. Michael Hawley, then an assistant professor at the MIT Media Lab, explained that his team had tried to launch a web archiving project. "We thought it might be possible over a ten to twenty-day stretch to capture the entire content of the Web and put it in a little time capsule for future generations," Hawley explained, noting that they were ultimately stymied. The "growth rates of data on the Web" meant that Hawley's team was "no longer able to do that . . . the net [has] now grown past our ability to suck it back in."[84]

Similarly, Rothenberg noted the problems around how to decide just what to select, presaging a problem that would vex archivists over the coming decades. If a document linked out to sixteen other sources, for example, would all sixteen of those need to be preserved? And what about the ones that those in turn link to? "You can think of the web as one huge interlinked connection of documents, you could think of it as a single document if you wanted to," Rothenberg explained, "it's dynamic, it's changing every moment, people are adding things to it, modifying things to it." Peter Lyman, university librarian at the University of California, Berkeley, added to this, noting that the underlying fundamental issues with creating a digital library were various. As Lyman asked, how best could web archiving happen absent government or even a centralized funding source? How could one develop "something more structured, something that thinks long-term about issues such as preservation, access, quality of information, quality of access?" As would be expected in 1997, the questions were many and the answers few.

Into the Future was well received and widely reviewed. "How fast do archivists have to run to stay in the same place?" wrote Paul Wallich in a review of the film for *Scientific American*, highlighting the challenges raised by web archiving and concluding that "where the Web was once a map for finding useful information in the 'real world,' it is now a territory where that information, ever changing, resides."[85] The American Library Association recommended the film "for all librarians and informed lay-persons."[86] In the *Information Management Journal*, Juanita Skillman reviewed the film as "a strong wake-up call we all need."[87] Crucially, writing in the American Historical Association's professional magazine *Perspectives on History*, Pillarisetti Sudhir gave an in-depth laudatory review that explored how the film contributed "a feeling of unease with the present fascination with electronic recording."[88] The wide range of voices who reviewed the film suggested that *Into the Future* made a significant intervention in how many of these communities viewed the web.

What was perhaps the most telling about *Into the Future* was its reception by professional librarians and archivists. It is often difficult to understand the degree to which something is widespread knowledge across a profession. By 1997, it was clear that digital preservation was on the radar of many information professionals. "While those of us within the library and information professions may well learn from this presentation," noted John Budd in a library journal review of the documentary, "it may be most effective to demonstrate clearly to college and university administrators, library boards, and others in decision-making positions the need for clear thinking with regard to information technology."[89] Sherelyn Ogden echoed this in *Library Quarterly*, as she noted that, while "the points made in the film will be familiar to most librarians and archivists, they are made very well and bear repeating."[90] This was a significant shift worth underscoring. If in 1990 the term "digital preservation" did not formally exist to cohere the range of activities that were beginning to take shape in that field, followed by the Task Force on Archiving of Digital Information (1994–1996), which introduced many to this new world, by 1997 digital preservation was sufficiently commonplace that not only could a PBS documentary be made about it but that it would be seen as more or less assumed professional knowledge in its reception. This was a rapid development. The field had come a long way.

Into the Future also spurred further conversation. Writing the next year in 1998, Margaret MacLean and Ben H. Davis reviewed the film: "Even handled as it is in a low-key fashion, it is a sobering experience to witness prophets such as these acknowledging the enormity and seriousness of the problem—the lack of agreement, tools, or standards for ensuring the survival of cultural heritage in digital form."[91] But what to do? A gathering held that year would try to bring all these disparate strands together.

"This Is No Way to Run a Civilization": The Conversation Comes Together at Time & Bits

As 1998 dawned, discrete threads and conversations were happening across the nascent world of digital preservation. Librarians and archivists had wrapped up the Task Force on Archiving of Digital Information, Kahle had started the Internet Archive, and technologists such as Myhrvold and Sterling had brought the issue to their respective communities. Another community would come together with these groups in a large conversation: thinkers concerned with the long-term future of humanity. These new conversations would combine the tangible questions of electronic artists with the long-term philosophical thinking of the "Long Now." This would happen at the Time & Bits conference, held at the Getty Art Institute in February 1998. This would be one of the last wide-ranging and high-profile gatherings on the topic for years to come.

It was appropriate that the conference was held at the Getty. Artists had been early web pioneers, drawing on the affordances of hypertext and new media to create rich online art and exhibitions. Despite web art's vulnerability to digital loss, however, preservation concerns were largely absent from much of the early commentary on net art—perhaps a result of it being such a new medium.[92] Yet by 1998, it was a growing concern as seen in the Getty's decision to host this event. A year later in 1999, the New York City arts organization Rhizome would establish "ArtBase," an expansive online archive for new media art.[93] ArtBase would later emerge as a significant player in the digital preservation space. Early pioneering web-based art was at risk of disappearance.

"Time and Bits: Managing Digital Continuity" brought together an eclectic group of individuals. Some will be familiar names: Kahle, Sterling, Lyman. Others were new to the conversation: Stewart Brand, founder of the *Whole Earth Catalog*, who was then raising awareness around the digital dark age; the musician and innovator Brian Eno; *Wired* editor Kevin Kelly; virtual reality pioneer Jaron Lanier; journalist John Heilemann from the *New Yorker;* Broderbund Software CEO Doug Carlston; futurist Paul Saffo; and digital archiving specialist Howard Besser.[94] Apart from a public session at the end of the two-and-half-day event, the participants met in a closed session. Their conversations built upon earlier private online discussions held before the meeting.

The event was sponsored by the Long Now Foundation, a group with complementary aims to those of the digital preservation community. The Foundation grew out of a project that had been bubbling around Danny Hillis's head—the parallel computing guru who, as we will see in chapter 3, first hired Kahle out of MIT—since the mid-1980s: the Clock of the Long Now, or the 10,000-year clock. Hillis's idea was to "build a clock that ticks once a year. The century hand advances once every 100 years, and the cuckoo comes out on the millennium."[95] To build such a clock would require long-term thinking. The clock idea would bring people together to examine the different aspects of such a project: Stewart Brand thought about the organization that would sustain it (which would later become the Long Now Foundation), and Eno coined the name itself. A 10,000-year clock would be both a social and technical challenge, as Hillis noted: "Ten thousand years—the life span I hope for the clock—is about as long as the history of human technology. We have fragments of pots that old. Geologically, it's a blink of an eye. When you start thinking about building something that lasts that long, the real problem is not decay and corrosion, or even the power source. The real problem is people. If something becomes unimportant to people, it gets scrapped for parts; if it becomes important, it turns into a symbol and must eventually be destroyed."[96] The Long Now Foundation was established in 1996 to develop two projects: the 10,000-year clock as well as its "Library" project, which would develop two main tools. First, a "Rosetta Disk" which, rather than being an optical disk, would contain over 13,000 pages in 1,500 languages "microscopically etched and then electroformed in solid nickel." Pages could then be read through a "microscope at 650X as clearly as you would from print in a book."[97] Second, the "Long Server" project, which was the "the over-arching program for Long Now's digital continuity software projects."[98]

The larger philosophy behind Long Now centered on the idea of civilization as having stretched back some 10,000 years. Inspired by ideas of extending the idea of "now"—the concept of which might refer to timescales as various as this exact moment to the exact week one is living in—to a 200-year time horizon. As digital humanist and librarian Bethany Nowviskie has evocatively explained, the Long Now expounds a "puckishly provocative *optimism* in everything they do," as opposed to complementary yet more pessimistic projects, like the Dark Mountain Project, that look toward the end of the world.[99] Indeed, Brand's account of the digital dark age was among the first to best grasp the utopian prospect of big historical data. "If raw data can be kept accessible as well as stored," Brand enthusiastically explained in 1999, "history will become a different discipline, closer to a science, because it can use marketers' data-mining techniques to detect patterns hidden in the data."[100] This was the subversion of the dark age concept. What if a golden age of memory was instead upon us?

This kind of big conceptual thinking was characteristic of Time & Bits. The whole idea, as articulated by the conference organizers, was to "do some 'out of the academy' thinking."[101] Participants started by sharing problem statements, extensively discussing them, watching *Into the Future*, and then assembling for a closing panel discussion. The most pivotal background paper, prepared by Lyman and Besser, outlined the problem that society and their small gathering alike faced: "our digital cultural heritage is disappearing, almost as fast as it is recorded."[102] Their essay succinctly summarized where things were in the field, notably in terms of networked information (must you follow hyperlinks to preserve a document?), strategies for preservation including how selective a collector should be, as well as how lessons from earlier format issues such as decaying acid paper could apply to this new problem. The discussions were recorded by organizers and subsequently published by the Getty Institute. Brand also used the event as a basis of a chapter in his 1999 *The Clock of the Long Now.*

The first task of the event, of course, was to understand the problem's scope. What was data? "Anything that can be copied," argued *Wired* editor Kelly. They then quickly, as a group, declared that, if possible, information should be saved in its entirety. As MacLean wrote: "The group agreed that it would be preposterous to propose any kind of selection criteria on what information should be saved. Many good reasons were cited, including the view that no one has enough wisdom to know which data might be valuable from another perspective in the future, particularly when you consider the analysis of large amounts of seemingly useless data which might have important information en masse."[103] A basic goal was thus established: try to preserve everything that can be copied. Brand would echo this, noting that one had to try to preserve everything as "*you never know* what will be treasured later."[104] Here the "puckishly provocative optimism" of the Long Now (per Nowviskie) was apparent.[105] To make this point, Hillis brought a small replica of the Rosetta Stone to show three things. These were the "impossibility of predicting future importance," how losing something made it last over the long term (the discovery of the Rosetta Stone ironically made its long-term preservation less likely, he argued), and when it was discovered, it was "immediately recognized as something important."[106]

The web presented an increasingly difficult problem but also a promising opportunity for future historians. Brand later noted that with the web, "preservation goes fractal: infinitely branched instead of centralized. Yet this leaves the question, Is the Net itself profoundly robust and immortal, or is it the most ephemeral digital artifact of all?"[107] Optimistically, participants at Time & Bits began to focus on the web's potential robustness—an inversion of the digital dark age. Could the web leverage collective wisdom to help preserve objects? Lanier gave an example of how online aficionados had been saving old video games, but of course, the propensity for links to break and servers to flicker offline made a strong case for fragility and thus the argument for active preservation.[108]

Bringing together the concept of the Long Now with the need to have preservation *and* access, Kahle proposed to the group an idea of "breaking the preservation into two parts: into access-oriented media—easier to read and write— and long-term, say 10,000 years, like a time capsule that's really hard to write."[109] Yet, would even making all these copies and preserving them help when one thought along the lines of millennia rather than years? Lanier thoughtfully observed that so many assumptions were embedded in contemporary computing discourse: "When undergraduates come into computer science departments, they are told about the idea of a 'file' as it were a fact of nature, as if it were as fundamental and immutable as a proton. [Lanier] likes to point out that the idea of a file has become locked in place as an idea because of its use in systems. In fact, it is a human invention that resulted from decisions that might easily have gone another way. The first version of the Macintosh didn't even have files."[110] These were big questions. How could one preserve culture for millennia, move beyond underlying assumptions, to move beyond a pragmatic discussion of emulation, migration, and file fixity toward the philosophical questions around what it is that a society of civilization leaves behind? These diverse comments came together at the final presentations.

"This is no way to run a civilization," declared Brand in his opening remarks at the concluding public forum. "Brewster Kahle pointed out that one of the peculiar things about the 'Net is that it has no memory. It's as if it's now the main event for civilization? We've made our digital bet. Civilization now happens digitally. And it has no memory."[111] By memory, Kahle referred to the ephemerality of these digital sources, as opposed to papyrus and other print materials. Next to the stage where Brand hosted the event and called up all of the speakers to discuss their thoughts on the issue, stood sculptor Alan Rath's art installation "World Wide Web, 1997: 2 Terabytes in 63 Inches"; a tower of four rack-mounted CRT monitors, which would display pages from "500,000 sites gathered and stored by Alexa Internet." Archived web pages flickered behind the on-stage panelists. To introduce what was at stake, a screening of *Into the Future* was held immediately before the discussion.

As befitted the only active web archivist at the gathering (and one of the few in the world), Kahle was optimistic. "The first reaction," Kahle explained, "tends to be, 'Oh my God, it's all going away.'" For an example of that perspective, he highlighted *Into the Future*. Yet Kahle offered more hope than the documentary did. "There's also this twinkle that comes up, which is, now that this is in digital form, we can do fantastic new things that we were never able to do before, in terms of making sense of it all, collecting it, data mining it, moving it all forward." Kahle argued that they needed to "try to preserve it all" but warned that "if we don't adapt soon, we'll go through a dark period."[112] Hillis, Kahle's former colleague from his supercomputing days, then raised apocalyptic warnings of a "digital gap." "The historians of the future will look back and there will actually be a little period of history around now where they really won't have the information," Hillis warned, emphasizing that it was "really the first time that the basic creations of a civilization are being stored on media that won't last a lifetime."[113]

While the conversation was intellectually diverse, the attendees were not. Apart from MacLean, one of the organizers, all the active participants were men. Hedstrom was a notable absence from the stage, given her work in this field, as were many other voices that could have been drawn from the field that we have seen in this chapter.

The other omission was that of future users: where were the historians? And, for that matter, where have they been up until this point in the chapter? Across the Dead Media Project, Documenting the Digital Age, *Into the Future,* and Time & Bits, historians were mythical constructs: imagined in an idyllic future state, poring over future websites, rather than the actual professionals working at that contemporary moment. Surely historians would be interested in this material, it was largely assumed, and would be able to enjoy these fruits of abundance. Discussed in theory, they were rarely present. Of course, it was true that historians would need to know about the present in the future, and to do so they would need access to these kinds of records. But what would that access look like? Just where were the historians? It turns out that the timing was simply not ideal, as by the mid-1990s, mainstream historians were at the nadir of their engagement with technology.

Historians in the Digital Wilderness

Despite the professional caricature of historians being uninterested in technology, historians have a long track record of engaging with digital information, including critical questions around digital preservation and access.[114] Amongst the historical professionals, these early encounters were driven by a wave of digitally assisted historians in the 1960s and 1970s who had then been relatively central to the historical profession. Historian Robert Swierenga posited in a 1970 retrospective of the "computerized research" field that historians were then in the midst of a third wave of computational historians, following 1930s punch card users who sought to tabulate quantitative data such as land mortgage information, to the second wave of 1950s or 1960s scholars who used sophisticated machines to understand historical demography.[115] The interdisciplinary journal *Computers and the Humanities* was established in 1966, and indeed, by 1970 the prospects for "computer-assisted historical projects" augured a wholescale transformation of the discipline.[116] Would all historians become digital? The peak of this conversation happened in the 1970s and early 1980s. When digital archivists and librarians convened the first conversations on electronic records in the 1960s and 1970s, historians were prominent participants. It would make their absence during the web age even more notable.

An early encounter was the 1968 Conference on the National Archives and Statistical Research, held at the National Archives of the United States. Occurring during quantitative history's apex, this conference convened historians, archivists, sociologists, and other scholars who were concerned not only with the use of digital records for historical research but also how to ensure that contemporary digital records would be preserved for the next generation. The issue at hand was the deluge of information accumulating at the archives. The "records since World War I far exceed in volume all earlier records in the National Archives," noted Meyer H. Fishbein, head of Records Appraisal at the National Archives.[117] James B. Rhoads, Archivist of the United States, warned of the "vast quantities of data" being accumulated by his institution.[118] These dire predictions anticipated those that would come in the 1990s in both substance and style. Economists and demographers made strong cases for the need to preserve information in a machine-readable format, and they even raised the prospect for remote access to archival holdings.

If historians had been strongly represented in these conversations during the 1970s, however, this changed by the mid-1980s and 1990s. Mainstream historians retreated from quantitative, and thus computational, work. The controversy around the quantitative history of slavery *Time on the Cross* as well as arguable hubristic overreach (Le Roy Ladurie's 1968 claim that "the historian of tomorrow will be a programmer, or he will not exist") combined, with other factors, to lead to its relatively rapid decline.[119] As Edward Higgs recalled, there emerged hostility amongst historians toward these quantitative, digital practitioners. "They were coming out with this sort of stuff saying, you know what, all historians are essentially out of date," Higgs recalled. "And all this stuff got up people's noses, something dreadful." Besides, the cultural turn became prominent, leaving digital and quantitative historians sidelined and—a term no historian wants to be associated with—niche.[120] Postmodernism was dominating conversations. As the stability of primary sources and their meanings was disintegrating, perhaps there was less emphasis paid toward new kinds of records.[121]

While what would later become the digital humanities was percolating in the background throughout this period, it would not compensate for this shift that was underway. It is somewhat surprising that there was not more overlap between the nascent field of the digital humanities and quantitative historians, given a shared interest in computers. This perhaps stemmed from the emphasis in the digital humanities toward computational literary studies and its attendant emphasis on marking up documents for analysis, an approach that did not easily scale with electronic records. In any case, the shift for historians away from quantitative methods and toward more traditional social history meant that historians became increasingly disconnected from pathbreaking digital projects.[122] As networked communication arose in the 1980s, and the web by the early 1990s, the timing was terrible.

Those historians working in the world of electronic records were among the first to notice the change that was happening around them, as their archivist and librarian colleagues became alarmed about these new records. Higgs recalls worrying about the longevity of email records, in a context where the Public Record Office would become involved only twenty-five years after the creation of a document: "Now, people are not going to be hanging on to things like emails for 25 years. And that was the sort of thing that was concerning me and other people at that time."[123]

This concern began to percolate into a little bit of historical scholarship. One of the first explorations of historians and born-digital records came in a groundbreaking special issue of the journal *History and Computing* on electronic records. Edited by R. J. Morris, the noted social historian and expert on class formation in nineteenth-century England, the issue looked ahead to what the social and economic records of the 1990s would look like to a historian in fifty or sixty years. Many of the social and economic historians used government records and thus were attuned to the shift happening within governments as these institutions shifted toward electronic records. This special issue, "Back to the Future: Historians and the Electronically Created Record," appeared in 1992. The editor's introduction explained the problem at hand: "When did you first hand somebody a text on a disk or send it by e-mail? Almost certainly this did not represent a sharp break in continuity in your practice as a historian. The implications of the changes taking place in the nature of the historical record needs to be assessed by historians even before that process of change is complete."[124] The editorial expressed fears that historians might look back to the two centuries between

1750 and 1950 as the "golden age of paper based history, with few telephones and almost no computers."[125] The problem of digital preservation was well explained from the historian's point of view, as the editorial pondered whether a policy historian, for example, should "at least have the ability to experience the data as it was experienced by the historical actors at the time?" This might require the use of "preservation or reconstruction of the main frames of the 1960s, or . . . software which simulates SPSS version one running on a KDF 9."[126] Similarly, document types were changing, leading to questions around just what a document was and, relatedly, what should be preserved. For 1992, this was remarkable language in a historical journal. Yet of the four articles, only one was written by a historian. The editorial lamented this absence, noting that "it is clear that this debate will remain incomplete without a great participation from practising historians."[127]

The historical contribution, "Virtual Records and Real History," by Ronald W. Zweig was an important one. Zweig was a political and diplomatic historian, bringing a different perspective than an economic or social historian exploring tabular data. As Zweig noted, political historians until the 1990s had more or less not needed to consider electronic records. "This situation is quickly changing," he argued, "as the first machine-readable textual records deposited in archives are being opened to research."[128] This would bring new challenges. The "guardianship of office records" was shifting toward IT personnel, who "are not known for sentimentality or their interest in records that they have never seen or handled," a problem compounded by documents that could "contain links and pointers to many other (interlinked) files of 'documents' so that the hypertext links are part of the information that the document contains."[129] Zweig expressed these worries but ultimately remained optimistic, noting that having digitized documents would open new frontiers. "Computerized records will make it possible to use sophisticated search and retrieval techniques," he prophesied. While mere keyword searching would produce too many results, "if they are combined with an understanding of linguistic equivalences, proximity and Boolean searches, and other techniques used in text retrieval, it will be possible to control the results of a search and to improve its quality."[130] This was an early contribution to what would later become the field of computational history.

The following year, in 1993, the edited collection *Electronic Information Resources and Historians* was published. Coedited by Seamus Ross of the British Academy, and Higgs, the collection grew out of a June 1993 conference.[131] Hedstrom recalled the event as opening "the door to a whole bunch of international things" critical to building community and fostering international engagement.[132]

Echoing themes from *History and Computing*'s special issue, Ross opened the collection with the thoughtful "Historians, Machine-Readable Information, and the Past's Future." "Awareness among historians of the changing character of contemporary information resources is limited," Ross noted. Paper records were giving way to electronic ones not only within government but also in commercial operations and even in the consumer realm. Ross was hopeful: "The sheer quantity, diversity, and rich quality of the electronic information resources . . . would seem to indicate that the preservation of the information in electronic form could provide historians with a better opportunity to understand our period than the paper records alone could ever do."[133] His vision was prescient. This "age of electronic records," as Ross understood it, could swamp "future historians with vast amounts of digital information [that might] impede their research as they attempt to navigate through it."[134]

These fears were echoed by historians contributing to the edited collection. Kevin Schürer, a demographic historian who was also assistant director of the British Economic and Social Research Council Data Archive, noted that the long tradition of historians and computers needed to be considered. "Consequently, given current trends in computer-usage, surely it is not all some pervasive technophobia that has caused historians to start sounding the alarm bells in warning," Schürer wrote. He further encouraged historians to "learn the skills required or suffer the consequences," but he also noted that this meant more than learning to code.[135] What would be needed was an approach to understanding computing that would be more akin to paleography. Perhaps even "technological advances allowing the 'reconstruction' of otherwise obsolete software" would be possible, opening the door for a technological solution akin to DNA analysis or radio-carbon dating in other fields.[136] This utopian approach to technology stood opposite the apocalyptic rhetoric of the digital dark age. Perhaps a golden age could dawn after all.

Personal electronic communication was still young in 1993. Accordingly, most of the discussions focused on government and commercial records. In a companion piece, however, Schürer raised the prospect of "the diarist, novelist or would-be intellectual sitting at home with [their] word processor." How could a biographer understand them if only the final product was deposited in an archive?[137] Other issues including hypertext and context complicated the matter further, and Schürer's stressing of the context of its creation was an important one.

Morris offset the optimism with a less sanguine perspective: "Last week (June 1993) I brought home a letter from my daughter. It had been sent by e-mail, transferred to a 3.5″ floppy disk and as a source of information was useless without specific software and hardware. The medium was no longer the message. Access needed a technologically sophisticated method of intervention. It was no longer enough just to know how to read."[138] Email was key. "The age of the network," Morris noted, would be "by far the most imposing of the problems faced . . . it is not clear we even have the intellectual concepts needed to talk about the issues we faced. The meaning of simple ideas like document, text and context, of provenance and sequence fall slowly and inelegantly apart."[139] Answers were not yet there but could be found through historians becoming aware of contemporary information issues, archivists working with institutions at the moment of record creation rather than thirty years in the future, and more attention to the internet.[140]

By 1994, then, a small number of historians and information scholars had clearly realized and articulated how important preserving web and network-based material would be for future research. They were forward looking. At the time, the web's dominance was not assured. Few were on the web, and competing internet protocols like WAIS or Gopher could still have eclipsed the web as the main way in which people would navigate the global network.

In April 1994, a conference hosted in Hampshire, United Kingdom, tackled the impact that computer networking would have on the humanities. It was a wide-ranging event, addressing topics as varied as preservation, access, digitization, electronic publishing, and organizational impacts. In his conference introduction, Seamus Ross introduced problems of digital archiving and preservation in one of the first-recorded reflections on the difficulties of web archiving to come: "How will networked communications and scholarship be archived? Who should have access to the archive? What levels of documentation should be retained and how should it be generated? What standards of data encoding, compression, and storage media should be used? Who will finance the preservation? What criteria for selection will be used? . . . Are email messages more akin to oral communication than textual sources?"[141] Other presentations contemplated potential solutions to these overarching questions. Sir Anthony Kenny, chairman of the British Library's Board, presciently noted in his keynote address the need to expand legal deposit regimes to encompass electronic material.[142] Hedstrom later stressed the importance of archivists thinking expansively: "As more individuals, informal work groups, and 'virtual' communities use networks to communicate, carry on discussions, and conduct business, archivists will need to understand these forms of communication as well as they understand the use of electronic systems in more traditional organizations."[143] Yet historians were still absent. Looking backward in a 1998 essay, Ross would accurately note that "awareness among historians of the changing character of contemporary information resources has until very recently been limited."[144] Indeed, after the initial flurry around *History and Computing*, historians seemed to disappear from the conversation as quickly as they appeared.

This rapid crescendo and subsequent wane in the conversation amongst historians around electronic records, primarily in the years between 1993 and 1995, is a bit surprising. Higgs ruminated on why it might have been the case that scholars tended to "come together, [do] a bit of networking, and then [they] tend to dissipate and people drift off into other things." He wondered if that was perhaps a combination of boredom, but also more importantly, the niche nature of this field of work. "I wonder if it's such a niche thing [that] people didn't really get promoted?" Higgs speculated to me, adding that "so it didn't fit into intellectual structures, and it didn't necessarily fit into career paths . . . [history] is a profession, and it is a career structure. And, you know, you don't get very far from being a niche player."[145]

Given the niche nature of this scholarship, for many historians these issues would only come to the forefront with American digital historian Roy Rosenzweig's June 2003 *American Historical Review* article "Scarcity or Abundance? Preserving the Past in a Digital Era." Given Rosenzweig's importance in the field, it is worth exploring his approach in some depth. Within the North American, English-language historical profession, the *American Historical Review* is the top-tier flagship journal, with articles enjoying wide professional readership. For many historians, Rosenzweig's article would be their introduction to the conceptual flood of born-digital resources that accompanied the web, as well as an introduction to the broader transformation of electronic records.[146] The *American Historical Review* both published Rosenzweig's article and—due to the article's foreseen significance—hosted an online discussion where readers could discuss the article with him and amongst themselves.

Rosenzweig's significance to the digital history field more generally and born-digital records and historical scholarship more specifically is indisputable. Rosenzweig, an urban and American historian at the forefront both of scholarly fields and of ways to leverage technology to reach new and expanding historical audiences, founded the George Mason University's Center for History and New Media (CHNM). CHNM was the pioneering home of much of digital history's new wave of scholars in the late 1990s and early 2000s. Established in 1994, CHNM has the mission of using "digital media and computer technology to democratize history: to incorporate multiple voices, reach diverse audiences, and encourage popular participation in presenting and preserving the past."[147] It remains a significant hub of activity today: developing the citation manager system Zotero, pioneering new publishing platforms, and training and fostering an entire generation of digital historians. Although Rosenzweig died in 2007, his legacy lives on in the now-renamed Roy Rosenzweig Center for History and New Media.[148]

"Scarcity or Abundance" was Rosenzweig's contribution to the digital records field. He introduced the problem, summarizing the debates and discussions that had taken place, from archival conversations, Time & Bits, and *Into the Future,* before bemoaning the lack of interest from historians. Rosenzweig posited that, in part, the "detachment stems from the assumption that these are 'technical' problems, which are outside the purview of scholars in the humanities and social sciences. Yet the more important and difficult issues about digital preservation are social, cultural, economic, political, and legal—issues that humanists should excel at."[149] The traditional archival system would break down, he worried, especially in private collections where "preservation cannot begin twenty-five years after the fact." What if a writer's heirs found a "pile of unreadable 5¼" floppy disks with copies of letters and poems written in WordStar for the CP/M operating system or one of the more than fifty now-forgotten word-processing programs used in the late 1980s"?[150] Rosenzweig also provided an overview of the Internet Archive, noting its scope and the attending issues of long-term sustainability for a private archive supported at least in part by a philanthropic millionaire.[151] Despite these challenges, he was hopeful, noting that "Kahle's vision of cultural and historical abundance merges the traditional democratic vision of the public library with the resources of the research library and the national archive." After all, despite the scope of national and research libraries, on-site physical access was necessarily restricted only to those who could physically make it into the reading room.[152] He concluded by calling not only for better technical skills amongst historians, but crucially, for collaboration between archivists and historians. The time was *now:* "If the past is to have an abundant future," he noted, "historians need to act in the present."[153]

Accompanying the essay was the online discussion, which ran for the first two weeks of September 2003 on the *American Historical Review* website.[154] In one of the many sad ironies of digital preservation, the electronic discussion was lost and not integrated with the long-term journal record itself, but it was preserved by the Internet Archive.[155] While it was a small conversation of only a dozen authors and twenty-two posts, this both reflected the start of the academic teaching term (early September is not an ideal time for any discussion) as well as the lack of digital engagement by many historians.[156] Discussions included what could tangibly be done, the degree to which historians needed to acquire archival training, the need to reduce professional divides, and insightful points around the degree to which digital information loss was any different from the amount of information lost during all periods. An active participant, Rosenzweig provided tangible avenues to foster collaboration through professional organizations, and he stressed that, while earlier information might end up in an archive through neglect, "the difference with digital data is that it appears if we wait twenty-five years, it may be too late—we could have nothing rather than, say, 10 percent of the data."[157]

Historians played a complicated role when considering their impact on the field. An early generation of historians, such as R. J. Morris, helped initially bring the conversation forward on how they would be impacted. Yet then, between the early to mid-1990s and the early 2000s, historians disappeared from the scene. Part of this reflected the Anglo-American divide in the historical profession, perhaps, with much of the leading conversations taking place in the United Kingdom, where North American scholars were perhaps less involved. It also may reflect the early burst of digital history in North America, which was focused on public history topics, a force that would not fully join with web archiving until the terrorist attacks of September 2001.

Conclusion

By 1997, there was a widespread, if elite, cultural consensus in favor of digital preservation. From Sterling's Dead Media Project to Time & Bits to historians like Rosenzweig, there was an increasing understanding that digital preservation could not simply be the purview of governments and corporate librarians but rather would require large-scale, interdisciplinary collaboration to make possible. On the one hand, the digital dark age raised apocalyptic visions of a sundered historical record. Yet, for others, inspired by the spirit of technological utopianism, there arose a prospect that the digital dark age might give rise to a digital golden age: a historical record unlike any that the world had ever seen.

With some exceptions, the individuals and organizations discussed in this chapter represented a more elite perspective: technologists, historians, and media theorists. Individual users, facing a 404 error alone on their home computer or the prospect of a lost document, largely did not feature in these conversations. Yet perhaps on reflection that is unsurprising. Everyday users would experience digital loss on an individual basis. The thinkers and writers discussed in this chapter helped to reconceptualize the problem as a much larger social one. It would not just be the individual trauma of losing some personal documents, recipes, or correspondence, it would be the collective imperilment of our cultural history. Put this way, it was not a matter of an individual tragedy but a looming digital dark age.

But so far, much of what we have seen in this chapter was theoretical: talking about building capacity, significance, networks, and occasional calls for action that were not always balanced by concrete, deliverable steps. The next step was to build the actual memory infrastructure of the web that could preserve it at scale and then provide access to it. During Documenting the Digital Age in early 1997, held in San Francisco, participants were asked "What now?" The Internet Archive had already begun to answer that question.

Building the Universal Library

The Internet Archive

"The dream of organizing all knowledge has been discredited," pronounced Steve Steinberg in *Wired* magazine's May 1996 cover article. He continued: "[a]lthough a few have continued to dream of a universal library—Vannevar Bush, who described his Memex system in 1945; Ted Nelson, who has been working on Xanadu since the early 1970s—they are widely seen as laughable in our relative, postmodernist era."[1] Yet, Steinberg allowed that there were "hints" of "new systems for sorting and storing information" on the horizon. Could the web form the basis for a universal library?

A quarter century later, the Internet Archive has become the internet's universal library. The Internet Archive has collected trillions of individual webpages and over 145 petabytes (a petabyte is 1,000 terabytes) of unique content. Its holdings double in size roughly every two years, illustrating both the web's growth as well as the shift toward multimedia content. While many institutions around the world use a version of the Wayback Machine to provide access to archived web content, the "Wayback Machine" is often synonymous with the official version hosted by the Internet Archive at archive.org/web. Since the mid-2000s, the Internet Archive has also undertaken the comprehensive digitization of books, government documents, software manuals, warranties, junk mail, and beyond, as it seeks to achieve its audacious vision of "universal access to all knowledge."[2] Headquartered in a decommissioned Christian Science church in the west end of San Francisco, the Internet Archive has also emerged as a place to debate new social visions and how the web might reconnect with its early, decentralized origins.

How did this critical part of the web's memory structure develop? How did a small start-up grow to essentially define and create the web's memory landscape? Building on the cultural rhetoric and intellectual currents discussed in the previous chapter, Brewster Kahle built a sustainable web archiving organization. The Internet Archive understood the digital preservation problem as a social one, not an individual one, and accordingly developed an internet-wide approach to preservation. Such an approach struck a chord. Kahle's vision launched it, but the Internet Archive's large receptive audiences certainly helped to sustain it. The Internet Archive succeeded despite the fears of many. To understand how this happened, we need to explore how the project evolved from a small idiosyncratic institution to a long-term universal library.

The Founding of an Idea: From Thinking Machines to WAIS

The Presidio of San Francisco—built by the Spanish in 1776, handed over to Mexico after its independence, subsequently captured by the United States during the Spanish-American War, and operated as a military base until 1994—is a picturesque place. Near the Golden Gate Bridge, the Presidio found a post-military life as a downtown park and commercial district. In spring 1996, a small group of technology start-up workers moved into a small, two-story, white-shingled building that had been originally built as a residence and had later served as the military base's post office.[3] The building's past life as a hub of information exchange was fitting for the organization that leased the building: the for-profit Alexa Internet, along with its nonprofit arm (founded the same day), the Internet Archive.

Then 35 years old, Kahle had been involved in cutting-edge computational and information retrieval technology for over a decade. In 1978, Kahle enrolled at the Massachusetts Institute of Technology, where elements of the Internet Archive vision began to percolate in his mind. Kahle recalls being challenged in 1980 by a friend to "paint a portrait that is better because of the future. That's better because of your technology." Pacing back and forth between Boston and Cambridge on the Harvard Bridge, Kahle recalled, he decided that he would set out to "build the Library of Alexandria, version two. How do you make all the published works of humankind available to everybody, forever, for free? And can you make it so that if you [have] anyone who was curious enough to want to have access to these materials, that they could get access?"[4] While many undergraduate students have audacious visions, the difference was that Kahle's vision would actually take shape.

The Library of Alexandria is an apt metaphor, for it captures two main dimensions of the problem facing the centralized accumulation of knowledge. On the one hand, the Great Library of Alexandria—located in Alexandria at the mouth of the Nile River, which accumulated knowledge through both aggressive acquisitions as well as copying (or perhaps taking) scrolls on boats that docked in Alexandria's prodigious harbor—represents the utopian dream of having all the world's knowledge in one place. Kahle drew upon this utopian vision in his dream. Yet there is an earlier cautionary tale around the Library of Alexandria, as encapsulated by Enlightenment thinkers. It "represented an Icarian dream: the greatest library in the history of the world, and it went up in flames."[5] For the Library of Alexandria also suggested the importance of broadly distributing knowledge. The Library of Alexandria was lost, a victim of long-term neglect. Multiple copies of texts, broadly distributed, can safeguard knowledge. The Internet Archive would later heed this lesson as it began to host its data around the world.

In 1982, Kahle graduated from MIT and was recruited by Danny Hillis—who as an MIT doctoral student first met Kahle while seeking undergraduate volunteers—to become one of the first employees of Cambridge, Massachusetts–based Thinking Machines.[6] We have already encountered Hillis via the Long Now Foundation. Thinking Machines was a fascinating company, with a range of ambitions spanning artificial intelligence to the (linked) development of parallel supercomputing. The company was also situated at the heart of the MIT intellectual technology community. Chuckling in a self-deprecating fashion as he mentioned a series of luminaries who also worked with Thinking Machines, Kahle recalled conversations he had around this time with Stephen Wolfram (later founder of Wolfram Research) and Richard Feynman (the Nobel Laureate physicist), and their thought experiments around how one could store all this information in a computer:

> This was a time when we were building a supercomputer that had 32 MB of RAM, really. And we were building the biggest disk system . . . It had five gigabytes in it, right? So, this is the high end, multimillion dollar supercomputer of the mid-1980s. But we knew that it was going on a particular curve. So, we just stood out the curve and said, all right, when will this be doable? And it was, you know, a decade, a couple of decades. But it was, you know, three decades. And all the technology would be there![7]

Building a second version of the Library of Alexandria was essentially impossible with the technology of the 1980s. But Moore's Law of exponential growth within the computing field portended future opportunities.

All this data would be useless, however, unless one could understand and process data at scale. Enter the idea of a "thinking machine." While parallel computing as a concept had been around since the 1950s, Thinking Machines aimed to develop machines with up to one million processors that would allow the "process[ing of] vast amounts of data on a real time basis."[8] By 1991, Thinking Machines had convincingly demonstrated the superiority of parallel over serial computing architectures, with one of its computers being declared the "world's fastest computer" in one competition.[9] Despite technical success, Thinking Machines faced commercial struggles and by 1994 entered chapter 11 bankruptcy and layoffs.[10] By 1992, however, Kahle had left Thinking Machines to spin off a separate company based around providing commercial services for the Wide Area Information Server (WAIS) protocol.[11]

WAIS was an important part of the internet's early history. It existed alongside the web and services such as Gopher as a way for people to access the growing global network. As Kahle explained to the *New York Times* in 1991, the idea behind WAIS was to apply supercomputing to the "problem of 'information overload.'"[12] Similar to how Berners-Lee positioned the web as bringing order amidst information overload, WAIS had its origins as a way to internally provide "corporate memory" within Thinking Machines.[13] Both the web and WAIS were internal solutions to a problem that would have much broader social significance. The development of WAIS began in 1989, when Thinking Machines partnered with Dow Jones, Apple, and KPMG Peat Marwick to design a system for searching across multiple networked information sources using natural language queries. Kahle moved out to the Bay Area to work on this project, inspired by a friend's advice that you should "go someplace where people don't think you're crazy."[14] Each partner played a different role in the project: Apple developed the front end, Dow Jones provided data, and KPMG Peat Marwick both provided corporate data and conducted user testing.[15] The project idea was to have "personal, corporate and wide area information all woven into the same system," and it would be used by executives and other end users themselves (as Kahle noted, it wasn't going to be a "call your librarian or your assistant to do it for you" type of system).[16] It was essentially an early stab at weaving the disconnected internet together into a universal library, with one access point.

A WAIS user searched across diverse internet sources using plain-language phrases. For example, a WAIS search for "Apple Computing" presented aggregated results from news sites, stock markets, business reports, and any other databases that might be connected. WAIS used the Z39.50 protocol, which had been developed by research libraries to connect with each other to facilitate interlibrary loan and information exchange. While WAIS was free for users, the computational power required to drive searches suggested a potential role for Thinking Machines supercomputers as underlying infrastructure.[17] By the early 1990s, WAIS was significant within both the library and information retrieval community. Mary Lukanuski, a data collection librarian at the RAND Corporation, wrote about the prospect (and fears) that WAIS presented: "Imagine the capacity to search a variety of databases through one interface. Imagine searching in everyday language, without having to use Boolean operators. Imagine not logging in and out when changing databases. Imagine accessing text, sound, and images with the same interface. Before you say 'Not!'—wait. This is all being done within the Wide Areas Information Servers (WAIS) protocol."[18] Screenshots confirm the relative ease of searching using WAIS with its graphical interface. It is also worth underscoring that WAIS predated the web in its public availability (and for a time contended with the web to be one of the main ways that users would access the internet).[19]

Kahle left Thinking Machines in 1992 to establish WAIS Inc. in a picturesque mansion in Menlo Park, California. The bucolic surroundings presaged the Internet Archive's unique future locales. As Kahle noted to me, "if you're working in Suite 201 in some anonymous office park, you will start thinking like Suite 201 in some anonymous office park, you will just start thinking generically."[20] He was joined there by John During, who provided the business acumen for a standalone company.[21] Kahle had already left Massachusetts for Silicon Valley to work out of the Apple offices while part of Thinking Machines' WAIS project, and ultimately, felt that the supercomputing-focused parent company would not provide the necessary investment that WAIS needed.[22] The new independent company would be built on the freeware developed at Thinking Machines, largely allowing Kahle to sidestep intellectual property issues.

The surroundings also made sense in a more abstract, intellectual sense. WAIS, and the later Internet Archive, would very much be influenced by the "California Ideology." This idea, originally emerging out of the techno-utopianist visions of venues such as *Wired*, drew on a larger intellectual lineage stretching from earlier countercultural and communalist individuals and groups such as Stewart Brand and the Whole Earth Network. As historian Fred Turner has noted, this ideology was one of "tear[ing] down hierarchies, undermin[ing] the sorts of corporations and governments that had spawned them, and, in the hierarchies' place, creat[ing] a peer-to-peer, collaborative society, interlinked by invisible currents of energy and information."[23] This well describes both WAIS and the later Internet Archive.

Built on a freemium model, the company would merge freeware—the WAIS software was freely available so anybody could run a basic server and access the network—with value-added services.[24] This was an early version of a business model that is widespread in today's open-source software community. WAIS Inc.

aimed to help corporations make money but also to capitalize upon "a new type of literature coming from people unleashed from the constraints of traditional publishing; we believe anyone can be a publisher."[25]

This built on Kahle's professional experience with publishing and digital libraries. The company also planned to leverage artificial intelligence to improve the relevance of returned search results. "WAIS is like a research librarian who watches you read through a stack of information, taking notes on what you looked at first and set aside for future reference, and what information you threw away," Kahle explained to *Forbes*.[26] Within a year, WAIS Inc. was providing services to customers as varied as the Environmental Protection Agency, Rice University, and the Library of Congress.[27] In 1994, WAIS Inc. was running a profit.[28]

By mid-1994, WAIS had become one of the main ways to access the internet alongside the web and Gopher. That would soon change. The web was growing increasingly prominent as it became increasingly useful and accessible, particularly with the launch of the NCSA Mosaic browser. Seemingly overnight the ease of Mosaic made it clear that the web had "won" and was even becoming synonymous in daily vernacular with the internet itself. WAIS would need to adapt to this reality. The WAIS team was soon looking for ways to incorporate their service with the web. One solution was to market WAIS as the web's "index," a useful complement to web browsing. The connection between WAIS and the library community also continued. In summer 1994, Kahle was invited by the Librarian of Congress, James Billington, to help identify "the major technical issues that must be addressed before the electronic library can become a reality."[29]

Increasingly tight operating margins led Kahle and WAIS Inc. to seek a larger company to acquire them. The hope was that an infusion of funds could allow the company to further develop its products. WAIS Inc. approached a seemingly unlikely partner: America Online (AOL), the online behemoth that by the mid-1990s was faltering as it struggled to adapt to the open web. AOL offered a bridge to the internet of the time. AOL had been a "walled garden," a platform not fully connected to the broader internet. The writing was on the wall for these walled gardens as the web exploded in popularity. Accordingly, by 1994, AOL needed to develop a plan on how to integrate its services with the web and internet. AOL was especially attractive to Kahle, as AOL had adopted an innovative royalty model vis-à-vis publishers, and WAIS did not have a way to pay publishers who were using its platform. Conversely, AOL was interested as WAIS was helping companies and individuals to become internet publishers. As Kahle noted, "one of the cool things was that AOL had a business model where they shared revenue.

So, they made six dollars an hour and about 10 or 15 percent of that went to publishers."[30] A marriage seemed to make sense.

Kahle and WAIS Inc. pitched AOL in May 1994 for project "Lightning," a "fast-track technology that [would let] any AOL IP publish to the Internet with little effort." Subsequently in January 1995, Kahle wrote David Cole—president of AOL's Internet Services—to "figure out a way for AOL and WAIS to work together . . . [I] am leaning towards acquisition in the current discussion and I have thought about price range a bit."[31] The acquisition was announced in May 1995 for $15 million in stock.[32] Alas, the royalty model that had excited Kahle was also on its way out with AOL. The timing for this was poor. "They changed their business model to where they actually charge publishers to get to their eyeballs. And that was, so that's fundamentally, an advertising-based gimmick," Kahle lamented.[33]

WAIS Inc.'s experience as a wholly owned AOL subsidiary was a challenging one. It is worth briefly exploring, as it would color Kahle's later activities. During his AOL tenure, Kahle primarily consulted on the pivot that AOL would take toward the internet. Post-acquisition WAIS was largely neglected by AOL, as it struggled with a hiring freeze, neglect of its existing contracts, and eventually its absorption into AOL proper as a department.[34] Correspondence between Kahle and AOL senior management became heated. "This is not what either of us had in mind," wrote Kahle to AOL's CEO in August 1995. By September, Kahle was being told it was a distraction for him to keep his office, as it was "clear that WAIS's role, post-acquisition has not gone as you would have wished, everyone knows this since you tell them . . . a clean break is best for all involved."[35] WAIS Inc.'s existing commitments were also endangered. In one exchange, the White House wrote to AOL with alarm at the slow pace of upgrades.[36] While there was a short-lived plan to make WAIS Inc. independent again, this fell apart, and in February 1996 the locks on the WAIS Inc. space were changed and Kahle's network access terminated.[37] It was an inglorious end for a company that had been one of the early pioneers in accessing information on the internet.

During his final AOL days, Kahle began to formulate his next idea: an "internet archive." This was a natural evolution. WAIS had been concerned with indexing information, and this had entailed relationships with libraries and publishers. Indeed, WAIS had advanced the idea that anyone could "be a publisher." If anybody could be a publisher, who would make sure that these publications would be accessible in the future?

As early as November 1995, Kahle had written to AOL Technologies president Mike Connors about his archive idea. "I am interested in archiving large collections of Internet content . . . I am pursuing this on the side (in a non-profit way). . . . If this is not kosher, lets discuss."[38] After this conversation with Connors, Kahle wrote to his assistant, "I talked about my plans to create a non-profit to build a large scale Internet archive . . . I said I was doing it with my money and on the side. He said fine and said he was interested in hearing how it is going in the future."[39] Kahle began to build his internet archive.

Kahle had had this idea for a surprisingly long time. In April 1994, buried in a much longer magazine profile about Kahle was an almost parenthetical note that the entrepreneur had been thinking about "trying to archive some of the works that are created on the net that were never meant to be printed . . . there's cultural changes and sociological changes going on that are not even being documented, they're not being saved, and it's time to start addressing these issues."[40] Kahle had for years been driving access to information. Now he would help to preserve it. The vision of a Library of Alexandria, version two—first cooked up on a late night in Cambridge, Massachusetts, over a decade earlier—would come to fruition.

Building the Early Archive

The Internet Archive had its origins in late 1995 as a nonprofit project during Kahle's waning AOL days, funded with approximately $400,000 of his own money.[41] By January 1996, Kahle was arranging transfer of data and storage media. "Can I borrow that 500GB robot that you mentioned? It would allow me to go the next step without laying out serious bucks," Kahle asked Carl Malamud, who then ran the internet's first radio station.[42] In April 1996, the Internet Archive was formally incorporated as a nonprofit organization. Kahle would simultaneously cofound—on the same day—the for-profit Alexa Internet to work alongside the Archive.[43]

Yet even as Kahle was concluding his time at AOL, he was corresponding with the Smithsonian Institution in Washington, DC, about a web archiving project to preserve the 1996 American presidential election. This project began in February 1996, representing a significant digital preservation milestone in the United States. In 1996, it was not yet clear that the internet would be significant for the election. The 1992 presidential election had seen only a few discussion groups amongst "elite" Clinton supporters. By 1996, in comparison, web usage was more common but far from universal. Major party presidential candidates, some minor ones, "as well as nearly half of the Senate and about 15 percent of the House candidates" had websites that year.[44] Yet, as pioneering political web archive scholars Kirsten A. Foot and Steven M. Schneider noted, these were "mostly 'brochureware' or 'virtual billboards,' simply replicating in electronic form materials already distributed in print . . . Web campaigning was largely seen as a gimmick or, at best, an ancillary to 'real' campaigning."[45]

Regardless, the Smithsonian's National Museum of American History was interested in collecting this new medium. Kahle reached out to David Allison, chief curator of the "Information Age" exhibition, to see if they needed help with collecting election websites. Kahle recalls this early collaboration between the not-quite-yet-incorporated Internet Archive and the American government: "[The Smithsonian] had a room of presidential memorabilia. And [David Allison] said, you know, this Internet might become something. So, the first websites might be like the first bumper stickers! . . . We proposed with them to go and archive 1996 Presidential Election Websites . . . They said 'great!' And they gave us a piece of paper with the sunburst [logo] on it saying, yes, we're doing it with them. And they knew exactly why we needed that piece of paper."[46] Armed with official letterhead, emblazoned with the sunburst logo of the Smithsonian Institution—defense against lawyers in the untested waters of web archiving—the Internet Archive (a few weeks before its official incorporation) was ready to collect. Letters were sent to parties and political organizations in late February 1996. The first web archiving crawlers soon followed.[47]

On 7 March 1996, a press conference was held at the National Museum of American History, for media, the public, and—in what was almost certainly among the first uses of the concept—"WWW historians."[48] The press release was the first public mention of the Internet Archive. "The Smithsonian's National Museum of American History announces Web Archive '96, a program affiliated with the Internet Archive of San Francisco, California, to record and track the 1996 presidential election as it appears in documentation on the World Wide Web," it explained. The archive would be a collaboration between the Internet Archive, historians, and museum staff, who would transform "campaign ephemera into a permanent, searchable record."[49]

Despite the initial publicity blitz, few media stories were written about either this novel collaboration or Web Archive '96 itself. Given the nascent nature of the political web, perhaps it was ahead of its time. The focus, in part, was for the webpages to "take their rightful place among the memorabilia of the 1996 race in the museum's exhibit this winter," which did happen.[50] Kahle laughs as he remembers that they "put a computer on the table next to the first bumper stickers, you know, [next to] Rutherford B. Hayes's buttons."[51] This was a key step. Michele Kimpton and Jeff Ubois note that the "data was eventually incorporated into the Smithsonian's presidential archive," where it resides today.[52] Web Archive '96, however, was external validation that this work was important, a month before the Internet Archive incorporated.

Kahle had a decent financial windfall from WAIS's sale.[53] However, the scale of the web and the ensuing amount of work to preserve it would have failed if a sustainable structure and organization had not been developed. Survival required something beyond a hobby or academic project, requiring instead a sustainable corporate structure. Indeed, this structure allowed the Internet Archive to endure, unlike so many other technology start-ups.

Through March and April 1996, Kahle gave presentations at various events about his Internet Archive idea. His main presentation, "Archiving the Internet: Towards a Core Internet Service," represented the first public articulation of how the project might work (the slide deck was reused at several events). Let me focus on one presentation. Kahle delivered a comprehensive overview of his idea as a talk at the April Frontiers in Distributed Information Systems conference in Key West, Florida. Kahle's presentation answered four questions: "What is it? Who will care? Is it Possible? Why do we want to do it?"[54] As the first public articulation of the Internet Archive's concept, the questions and answers are worth exploring in depth.

First, what would the Internet Archive be? Its mission would be to "Gather, Archive, and Serve all public Internet information," including the web, Usenet, and Gopher. This was for three reasons: the reliability of having a backup of the web, the accountability that came with drawing on an official record, and crucially, the durability that such records offer researchers. Decades later, the Internet Archive's mission statement is philosophically similar.

Second, who would care? Prospective users included those who wanted reliable access to favorite sites, marketers seeking to gain a treasure trove of demographic information from a comprehensive archive, and finally entrepreneurs who could use all of this data as a "basis for new value-added services." Yet there was one more important group that Kahle envisioned using his archive: "scholars/historians" who would want to understand this "new medium." Kahle would be remarkably consistent (and prescient) concerning the value of the archived web to scholars over the coming years. In a 1998 *Chronicle of Higher Education* profile, Kahle argued in favor of preserving digital heritage in evocative terms. It was necessary, "or one day, digital anthropologists will wonder if we ever learned anything from the history of other inventions. Remember, nobody recorded television in the early days."[55]

Third, a big question: would it even be feasible to build an Internet Archive? The slide deck, given the technical audiences of these talks, delved deeply into this question. Kahle concluded that gathering 10TB (his estimation of the web's then-current size) would cost approximately $100,000 to purchase capital equipment (tape drives and servers) and another $10,000 annually in bandwidth costs. Not trivial money, but not beyond imagination. Should the right use cases and opportunities appear, an Internet Archive could underpin a sustainable business. These figures were, of course, just to preserve the information. Access, as Kahle noted, would "take [more] calculations and fresh ideas."[56]

Finally, why archive the web? The transformation of the web from ephemeral to permanent was key. So too were the downstream implications of data mining, a potential internet operating system, and beyond. All it would take was bandwidth, computational power, smarts, and gumption, concluded Kahle.[57] Throughout this period, too, Kahle was focused on the transformative nature of web publishing. In May 1996, Kahle explained to *Wired* that "[i]nformation retrieval is not about finding how much tannin there is in an apple, it's about letting everybody publish."[58] Access to knowledge was never an afterthought but a prime consideration from the Internet Archive's conceptualization.

With the importance and feasibility of the project sketched out, the final question was where the money would come from. The money Kahle earned from the sale of WAIS Inc. could certainly buttress the project for a few years given its reasonable costs. After all, Kahle was noted for his frugal lifestyle ("[a]lthough he and his family can afford to live like millionaires, they generally choose not to" claimed a 2007 *New York Times* profile).[59] Notably, however, the presentation did not just focus on historians or scholars but also on the business potential that one might find in such a treasure trove of archived data. In this would be the idea that would germinate in Kahle's other business idea: Alexa Internet.

Kahle pitched Alexa as the "Internet Companion company" to potential investors. It would form the sustainability model for the Internet Archive. Recognizing that web users were getting lost as they surfed the web, already encountering 404 errors, Alexa proposed an ever-present toolbar that would provide information about each page that a user visited. It would also provide access to archived web sites.[60]

The Toolbar worked via a proxy through which the user navigated the web. Today, for example, researchers may use a proxy to access research library materials from home. The proxy allowed a user to see both the page they were requesting as well as contextual information: the page's popularity, update frequency, geolocation, and—crucially—the other sites that users tended to explore. For example, a user could see that visitors to a page about the Mazda Miata convertible also visited Pontiac, BMW, or *Road & Track* magazine. In an era when search was still in its relative infancy, this service would help to bring order to the web.

To work, Alexa would need to comprehensively crawl and download the web. This archive of webpages could provide reliability for users to help them avoid the frustration of 404 errors, but just as importantly, the raw data of the internet would also be necessary to map the web's link structure. A local copy of the web would allow the computation of updates, background information, and beyond. In 1996 the speed of both computing and bandwidth meant that such an offline copy of web data would be critical to make any of that possible.

A team was in place to build Alexa, which would also simultaneously build the Internet Archive. Alexa would underpin the Internet Archive's sustainability. Kahle would be president. Bruce Gilliat, who had worked on WAIS with Kahle, would be general manager and VP of marketing. Gilliat would come with Kahle to his new enterprise to provide the necessary business chops and acumen. They had an aggressive timeline: to design and archive throughout the first three-quarters of 1996, with an eye to an early 1997 launch. Kahle concluded his investor pitch: "Your Internet Companion makes the Internet understandable and reliable."[61]

Building a Modern Library of Alexandria: Alexa Internet

Alexa's core product—the Alexa Toolbar, the realized Internet Companion—did not at a glance evoke its namesake, the Library of Alexandria. As Alexa launched alongside the Internet Archive, the media was curious about this unique corporate structure. From its inception, Alexa was somewhat overshadowed by its flashier nonprofit sibling. "The Internet's historian now wants to be its guide," Renee Deger noted in *ZDNet*. "The idea for Alexa grew out of another of Kahle's projects, the Internet Archive," explained Laurie Flynn in the *New York Times*.[62] As Kahle later recalled, once all of the data was flowing into the Internet Archive, "he realized how difficult finding things on line was becoming, with the number of Web sites doubling every six months even as other material falls into neglect. So he set about creating a Web search engine using the technology he developed to manage the massive amounts of data he was collecting for this quirky history project."[63]

A note on the timeline of the nonprofit Internet Archive and the for-profit Alexa Internet corporation is in order. Much of the early history of the Internet Archive is inexorably connected to Alexa. The two organizations were founded on the same day, with the Internet Archive serving as the long-term nonprofit archival backup for the material collected by the for-profit (and thus perhaps shorter-term) corporation. As Kahle later told me, he "wrote a contract into the soul of Alexa Internet: it would donate everything, all the data collected to the Internet Archive . . . starting in 1996, through the end of 2020, Alexa Internet every day donated data to the Internet Archive."[64] The two organizations were entwined: the Internet Archive playing the long game while Alexa was more in the vein of a traditional technology dot-com start-up.

The combination of for-profit Alexa Internet and nonprofit Internet Archive made for a complicated organizational structure. Alexa owned everything and employed everybody. But Kahle was aware of Alexa's inherent structural limitation as a for-profit company: "If there's one thing you know about companies, it's that they're not going to last, right? Companies self-destruct or they get acquired or they just, you know, they flub it. Right? . . . For-profit companies are like gasoline and just, you know, fire and it just goes WHOOSH. But if you want something that lasts, build a non-profit."[65] This bifurcated structure would work, especially as Alexa began to collect massive quantities of web data thanks to the Alexa Toolbar.

Announced in July 1997, the Toolbar connected to a web browser and tracked how a user navigated the web. It would then share aggregate information with other users. By collecting and analyzing user behavior at scale, Alexa learned where people who visited one site also visited. In a presearch web era, when finding content was difficult, there were clear opportunities to assembling this data. Users could visit their favorite board gaming site, for example, and other board gaming sites visited by people similar to them (based on their browsing history) would be suggested. As a corporate press release explained, "in essence, Alexa harnesses the collective wisdom of the Web by taking advantage of its most killer app—the people who use it."[66] Kahle put it more eloquently: "If you take a path through the woods, you benefit from the exploration people have done before you in finding the best way up the mountain or down to the lake."[67]

The Toolbar worked through regular crawls and harvests of the web, downloading material to analyze in a few ways. Approaches included text analysis ("looking for pages and sites with similar characteristics") and link analysis ("detecting patterns that point to particularly common or popular sites"). This information was coupled with browsing behavior to determine "anonymized surfing habits of its users: which sites they visit, how deep they go, how long they linger and where they go next."[68] This was a popular service for users confronting an overabundance of information and hyperlinks. The Toolbar received coverage across the mainstream media as well as technical media and websites. Indeed, as the pages navigated by Toolbar users formed the initial Internet Archive collection for at least its first two years, it is worth noting that the bias of users who would use the Toolbar (technology early adopters) skewed the Archive's early collection.[69]

Alexa Toolbar was advertisement-supported. The company hoped that aggregate user information would help appropriately target these advertisements and thus command a premium, like how Google's later AdWords could tailor advertisements based on its own voluminous data. Crucially, the Toolbar was always present and could thus always show advertisements to users.

Of course, the Toolbar did not just help users find information: it preserved data for future use. Tucked away in Alexa's first publicity release was a note that "Alexa is the only service that provides you with quick, automatic access to an archive of the entire public content of the Web. Now, instead of your browser displaying a '404 Not found' message, Alexa will retrieve a copy of the page from our archive."[70] The growing Alexa archive would be useful for four reasons. As Kahle noted, the first was to find "patterns in the Web to build a navigation service," or Alexa, and the second was to track "changes in the Web and its use to find areas of growing and shrinking interest." Yet the third and fourth were key: "offering historians and scholars an unprecedented collection of human voices" and "keeping a record of the birth of a new medium, and the dreams people have for it."[71] Amidst this initial wave of publicity, while most coverage focused on the navigational aspects of the Toolbar, John Markoff noted in the *New York Times* the ability to "retrieve 'dead' Web pages that have been captured [i.e., preserved] by the Internet Archive."[72] Soon, however, as the Toolbar receded into just being another useful aid for navigating the web, external attention began to focus on the significance of capturing web-based material for future use.

Disentangling the Internet Archive and Alexa Internet can get confusing. Alexa was the corporation that employed employees and provided space and equipment, whereas the Internet Archive was the nonprofit that would preserve collected material in the long term. While scholars have later questioned the use of "Archive" to describe the Internet Archive, in this context the term makes contextual sense: the Internet Archive was Alexa's archive.

The use of the term "archive," with its suggestions of stability and longevity, was an important counterweight to the reality that start-ups like Alexa were typically short-lived. If information was to be stewarded in perpetuity, a start-up would not be the right structure. As Kahle recalled in 2008, the dual Alexa / Internet Archive structure was a deliberate choice:

> The idea was that everything that Alexa ever collected would be donated to the Internet Archive. Over the long term, companies come and go. They usually don't last that long. But the great thing that was going on with the Internet wasn't the technology. That gets replaced. It's the information, and it's all the people. So we started collecting the World Wide Web and making services in a commercial com-

> pany, but donating all of the materials collected to a nonprofit that was designed to last the ages. It was very specifically designed to think through what happens after the commercial company is gone.[73]

Given that Alexa would later be acquired, this was prescient. Almost immediately, tape drives filled up at the Internet Archive, transferred there by Alexa. In practice, as the two organizations shared space and staff, these transfers were somewhat conceptual. Yet the long-term stewardship would fall to the nonprofit organization that had *rights* to the data. Given copyright's centrality to many decisions, this was fortuitous.

The excitement around saving the web attracted attention. As we saw in the previous chapter, Kahle was charging ahead even as much of the field was still discussing the contours of the problem. Vint Cerf, who was about to chair the Documenting the Digital Age conference, observed in a feature for *USA Today* that Kahle's "enthusiasm for ideas causes him to overlook practical considerations," such as "looming issues of copyright and privacy raised by copying Web pages without their creators' express permission." Kahle's response was that "by not worrying about the details he's able to do things others think are impossible." This engendered respect from Cerf, who noted that Kahle was "the kind of visionary who bears watching."[74]

For the initially modestly funded Alexa—it soon raised a million dollars in venture capital funding (the first million from Kahle, $300,000 from Bill Dunn; later this would be joined by $10 million from another private investor)—impressive but far from a blank check—archiving the web while also running the Toolbar service was challenging. Developing lean infrastructure would put both Alexa and, ultimately, the Internet Archive in good standing. Based in the newly on-the-market Presidio (the US Army had left only two years earlier), where Alexa was only the second commercial tenant, a scrappy, lean organization took shape. In the wooden four-bedroom house, "mismatched office furniture" housed computers and tape drives, crawling the web.[75] In keeping with Kahle's goal of not setting up a company in Suite 201 of a generic office park, but rather someplace more interesting, the Presidio fit the bill. Kahle joked that "Smokey the Bear was our landlord!"[76]

There were, of course, downsides to building Alexa and Archive in a wooden nineteenth-century house (the "little house on the prairie," as Kahle put it to me). Steven Schneider and Kirsten Foot, researchers who would later work with the Internet Archive and the Library of Congress during their 9/11 attack archiving, remembered an early visit in 2001 to the building. Foot laughed as she remembered "going to this wooden building room full of thousands of servers, right? . . . and then we're looking like 'Is there a fire protection [system] here? And what will happen when the sprinklers come on.' "[77] Yet, this attitude also made the Archive's ambition possible. With such relatively inexpensive infrastructure, data would begin to flow at a previously unimaginable pace. The Internet Archive began to scale.

Indeed, the sheer amount of data was overwhelming. By late 1997, Alexa had harvested some "600,000 Web sites and more than 135 million pages."[78] Tape after tape of web data was produced, subsequently being stored on quick tape-switching machines for automatic retrieval. Several observers thought that these looked like old jukeboxes, five 40GB tapes loaded into each machine.[79] As Kahle remembers:

> The only way to really cost-effectively store gigabytes of data was on tapes. But if you had tapes, then they stack up and you can't access them. So there are these tape robots and [laughs], these were miracle machines, right? They pull this tape out and go rrrrrrrr, then you put it in and then go and then spin it up, and three minutes later, they would seek the right place and you get your webpage back. It's like, "wow! Really slow." But it did work . . . I mean, this is a couple of thousand square foot "house in the prairie"–house. And it had this tape robot in it that went rrrrrr.[80]

It was certainly an evocative image for an institution that would end up collecting the lion's share of the world's digital cultural heritage. Schneider also remembered a laissez-faire attitude toward occasional data loss that perhaps made the operation possible. When it came to tape (or later hard drive) failures, he recalled that Kahle's attitude was "it doesn't matter. We'll just collect more . . . they were building machines as fast as they could. They were filling them up as fast as they could build them."[81] Kahle noted that this was a calculated decision. As he recalled, given his "limited budget for data storage, you had the choice of either keeping a duplicate to make sure it was secure, and then only collect 50% of what you could have, or you can risk losing less than 1% through data loss." To him, "possibly losing less than one percent of what you can capture is better than certainly missing fifty percent of what you can capture."[82] This spirit was, in many ways, probably the only way such a large undertaking could get off the ground. Perfect can be the enemy of the good, and given the web's ephemeral nature, speed was critical.

In addition to the technical infrastructure, these Alexa crawls also used a standardized file format to preserve web data: the ARC file. ARC files would become a de facto standard that combined archived web data with metadata. The

1.0 specifications, released in September 1996, represented the beginnings of a standards-driven approach to web preservation and access that would later help to facilitate a larger tools-building community.[83] In 2009, ARCs would be superseded by a slightly different format, the ISO-standardized WARC file.

What happened to Alexa? "Alexa, can you look this up for me?" Alexa is known today as Amazon's smart assistant, although the overlap in names may simply be a coincidence.[84] There was some foreshadowing of this in 1998, when Kahle described Alexa Toolbar in the *Washington Post* as "an Internet assistant, like having a really smart friend with you."[85]

Alexa Toolbar was widely adopted and made available for browsers including Netscape Navigator and Internet Explorer, but the ad-supported model ultimately did not lead to a fully sustainable business.[86] In April 1999, Amazon announced a high-profile acquisition of several technology firms, including Alexa. Alexa would help build Amazon's comparison shopping service.[87] This was a big purchase: $250 million in Amazon stock, and crucially, unlike the experience of WAIS within AOL, Alexa would remain largely independent within Amazon.[88] Importantly, Kahle "was allowed to continue with the altruistic side-project that he had specifically created Alexa for: the Internet Archive."[89] This is where the dual structure of Alexa and the Internet Archive paid dividends. The Internet Archive remained independent, and the agreement continued to transfer crawls from Alexa to the Internet Archive for over two decades after Amazon's purchase.

Kahle remained with Alexa for three more years before moving full-time to the Internet Archive. During that period, Alexa would receive more attention around its data-gathering practices. In January 2000, Alexa became the target of a US Federal Trade Commission investigation into the incidental collection of "personal information including names, postal address, phone numbers and e-mail addresses" during the harvesting of user activity on the web.[90] The investigation, and a subsequent class action lawsuit, led Alexa to settle an agreement, seeing it pay $40 to each user who had personal information in its database as well as a larger $100,000 donation to "Internet public policy programs and consumer groups."[91]

Throughout this period, Alexa continued to donate data to the Internet Archive. Indeed, it was only in 2003 and 2004 that the Internet Archive released its own independent web crawling program, Heritrix (which remains the backbone of web archiving operations today). Outputting files to the ARC format, later WARCs, Heritrix would give the Internet Archive the ability to do its own crawls without Alexa Internet. As we will see in chapter 4, this functionality was critical when the opportunity arose to do "contract" crawls for other organizations such as national libraries. Indeed, Heritrix itself was a product of a joint collaboration between the Internet Archive and developers at the Norwegian and Icelandic national libraries. Thanks to Heritrix, by 2008 the Internet Archive would begin collecting most of its own material without Alexa, although Alexa crawls continued to be donated to the Internet Archive until 2020. While Alexa remains today under the Amazon umbrella, it is worth remembering that for a time it provided the backbone of the world's web archiving.

In the Service of History: The Internet Archive and the Quest for Access

Despite Kahle's wealth, the early Internet Archive was technically modest. Its infrastructure consisted largely of "jukeboxes": those tape machines with five 40GB tapes apiece, forming the long-term memory for Alexa Internet.[92] An early visitor would have been excused had they had doubts that this modest beginning, in a small wooden house, would ultimately come to serve as a key part of the memory apparatus for a globe-spanning communications medium.

What is remarkable about the Internet Archive in hindsight is the degree to which Kahle and his organization maintained a consistent vision of why it was engaging in widespread web crawling: in the service of history. In 1996: "The Internet is millions of people every day discussing what's important to them, all available now, whereas before we had to collect diaries or personal letters after somebody was dead."[93] By early 1997, the vision was more audacious: "given that many of the world's greatest books, lyrics, images, and other artworks pass through these networks, an Internet Archive would, some say, be nothing less than the sum of all human knowledge."[94] There were nods toward the commercial potential of this data, but that was always secondary to the retention of data *regardless* of its contemporary value.

The question of access was a critical one from the Internet Archive's founding, particularly as it interacted with the issue of privacy. The first year of the Internet Archive's existence perhaps prompted the most public handwringing, as for the first time, observers contemplated the end of ephemerality and the specter of permanence. In July 1996, with the Archive only months old, David Berreby lingered on the idea of long-ago electronic posts living on for decontextualized posterity. Writing in *Slate*, he asked readers to "[t]hink about it: The most embarrassing e-mail you ever wrote, available to anyone curious enough to go looking . . . As we're encouraged to exult over the vast new volumes of information that are becoming easier and easier to capture, remember that the art of losing is also important to master."[95] Similarly, Dan Gillmor, writing in his widely syndicated technology column, recognized that, while he was "not crazy, either, about the prospect that every dumb thing I've said on-line will be saved," he did note that he was not "smart enough—and I don't have the time—to decide what's worth saving."[96] "Will the archive continue to offer revealing personal home pages long after an author has decided to pull the plug on them?" pondered Rajiv Chandrasekaran in the *Washington Post.* "It can't be the case that every time some one uses the Internet they have given up control of personal information absolutely and for all purposes," argued Marc Rotenberg, director of the Electronic Privacy Information Center.[97] John Markoff worried in the *New York Times* that the Archive could "follow the paths of millions of people and record their likes and dislikes, or even their communications . . . The more complete the Web's archive is, privacy rights advocates say, the deeper the Big Brother problem is."[98] This early coverage demonstrated an awareness of what would be a pressing problem. Given the Internet Archive's novelty, observers were asking fundamental questions about the status of it all.

The Internet Archive was not alone in making people uncomfortable throughout its first two years. New technology was changing the discoverability landscape. Steinberg's May 1996 *Wired* cover story on search made this point too, in the context of searching years of Usenet discussions. "It's possible, for example, to search on a person's name and find every message they have posted," Steinberg wrote, adding that "whether it's on comp.client-server or rec.arts.erotica. Using these tools, anyone can build a profile of a person's interests, based on where they post."[99] In other words, "[w]e can no longer depend on privacy through obscurity."[100] Around this time, online privacy was emerging as a problem. Books such as Carole A. Lane's *Naked in Cyberspace: How to Find Personal Information Online* comprehensively documented the sheer amount of stalking one could do online.[101] Lane herself noted in *USA Today* the problem inherent in preserving internet discussions, wondering whether a "student might post uncensored anecdotes or opinions, not realizing that somebody years later, perhaps a potential employer, might learn they had done things they were embarrassed by."[102] Yet despite this uncomfortable conversation, the Internet Archive was unabated in its work. Copyright and privacy problems were overwhelming, meaning that the initial emphasis needed to be on collecting. Thorny access issues would have to wait for later.

Would a different access paradigm work? One early idea was to somehow anonymize personal information. Kahle pondered in 1996 as to whether Internet Archive data could be "handled like census data—aggregate information is made public, but specific information about individuals is kept confidential."[103] From its very inception, and through to today, it was also clear that the Internet

Archive would honor any take-down request from a private individual to have their site removed from the collection.[104]

Access to the Internet Archive, however, soon adopted a replay paradigm thanks to the Alexa Toolbar. This presented two advantages that minimized the potential privacy impact. First, the Alexa Toolbar led to "privacy through obscurity." A user could access the archived webpage only if they already had the URL before encountering a 404. While this would make it harder for somebody to delete content they wanted to hide, it did not make historical information accessible to keyword searches. Second, data was on tape drives, which meant long recall times for all but the most accessed (and thus cached) data. Bruce Gilliat quipped that "getting a page could be done in a few seconds . . . or days later."[105] Even high-speed tape had retrieval times in the order of minutes as opposed to the seconds of hard drive retrieval. Popular sites could be put on the Internet Archive's smaller number of hard drives for quick retrieval—"if it takes people up to four minutes to find [archived copies of] Yahoo! on our system, they'll go away and never come back"—but most others would take minutes to be displayed.[106] One account observed that the wait could be even longer: "any one of the world's 135m web pages [could] be retrieved in about 15 minutes."[107] This slow, methodical access did not lend itself to widespread speculative browsing and, incidentally, would also help with alleviating privacy concerns in practice. Something more would be needed for researchers if they wanted to be able to use the Internet Archive for more extensive research.

The quest for better access would be facilitated thanks to two crucial developments. First, data access times were improved. The Internet Archive migrated content from tape to hard drives in 1999, dramatically reducing wait times. Hard drives would also increase reliability: the tape robots often broke down, and the format itself struggled. Kahle noted that the "tapes are often unreadable . . . the bits rot[ted]."[108] Second, the Internet Archive developed robust research access infrastructure. By 2000, approved researchers could remotely connect to the Internet Archive using a secure shell. Researchers could complete a form, discuss their specific project with Archive staff, and, if successful, gain access through this process.[109] It was a high barrier to access. Remotely accessing a server requires a high level of technical knowledge, beyond the abilities of most researchers in the humanities or social sciences. To truly realize a vision of a universal library, easier access would be necessary.

Enter the Wayback Machine, named after Mr. Peabody's time traveling machine from the Rocky & Bullwinkle cartoon. As early as 1997, Kahle mused to Markoff at the *New York Times* about "a useful service [that] would be an archive that would retrieve such links, perhaps for a fee. Mr. Kahle envisions a sort of 'dialable way-back machine.'"[110] With a wink, too, Markoff noted that if "all goes well, though, Mr. Kahle's Internet Archive won't cause nearly as much mischief as Mr. Peabody's way-back machine."[111] Trouble or not, Kahle's Wayback Machine would certainly make an impact when it appeared four years later.

The Wayback Machine was unveiled on 24 October 2001 at the University of California at Berkeley's Bancroft Library.[112] The Bancroft Library, a large special collections library with ornate historical decorations, gave a historic patina to the event. "You know, the walls and the ancient things," Kahle recalled, painting the picture of the day, "Berkeley knew exactly what they were doing and giving this new upstart idea that kind of venue for launching the Wayback Machine."[113] This was a historic event. As the *New York Times* reported:

> Mr. Kahle introduced his brainchild—named, yes, for the time machine used by the pedantic dog, Mr. Peabody, and his boy, Sherman, from the "Rocky and Bullwinkle" cartoons—with a flourish . . . He demonstrated it with a certified stunner: he pulled up a Web page from the White House Web site from Sept. 10, 1996, with a press release about President Clinton proclaiming the prevention of hijacking and terrorist attacks in the air a priority. Mr. Kahle said he had also used the system to read Web pages created by the Heaven's Gate suicide cult and to find a manual for a computer part that had been taken off of a company's Web site in 1998.[114]

Users loved the new Wayback Machine—perhaps too much so.

The Wayback Machine was popular beyond the Internet Archive's wildest expectations. The site remained intermittent for users for several months as the Internet Archive added more bandwidth. Kahle was surprised. He had "[t]hought it would be big, though he didn't realize just how big." As Kahle explained to *USA Today*, he figured, "It's just a library. People won't storm the doors of the library," but they did—at the rate of 50 to 100 requests a second—overwhelming the Internet Archive's capacity.[115] This attention raised fears that with added accessibility and visibility would come potential litigation. Over "the last five years the archive hasn't attracted much attention from the copyright cops," wrote Katharine Mieszkowski for *Salon*, "it's also true that its 100-terabyte holdings hadn't previously been just a click of the mouse away."[116] Ultimately, the surge of interest in the Internet Archive was exciting, vividly showing that better access would drive significant interest and engagement.

The Wayback Machine represented a genuine leap in accessibility, remaining the dominant paradigm of web archiving access up to today. Users would visit the Internet Archive website at archive.org, navigate to the Wayback Machine, and then enter the URL of the site they wanted to visit. For example, "nytimes.com" would then display a list of all the homepages of the *New York Times* available stretching back to 1996. A user would then select a date and then be able to view the older archived site. The major limitation, of course, was that the user needed to know the URL. But once users navigated to a page, they could "surf" as if they had traveled "way back" to that year.

The implications of this easy access to the archived web were dramatic. "The historical implications of the Wayback Machine are immense," argued Greg Notess in *Online Magazine* in 2002, "Historical researchers can now view significant portions of the Web as it existed at various times from 1996 to the present."[117] The Wayback Machine would remain largely as launched, with incremental improvements, until 2016, when the Internet Archive publicly launched its "site search" function, which allowed for keyword searching across website home pages.[118] In 2019, Internet Archive also added its "Changes" service, which allows users to pick two archived pages and highlight what had changed between them. Finally, the Internet Archive's "Save Page Now" launched in 2019 as well, allowing anybody to provide a URL to be immediately saved and added to the Internet Archive's collection.[119]

By the end of 2001, the Internet Archive was providing reliable, comprehensive access to its collections. With the web harvesting and access side of the organization established, after 2001 the Internet Archive began to focus on the much broader question of how to preserve diverse forms of information and ensure universal access to *all* knowledge. In part, this would require confronting an issue that hung over almost all of its activities: the ever-expanding power of American copyright and intellectual property regulation.

Copyright and the Internet Archive

If the story of web archiving needed an antagonist, it would be copyright law. The question of copyright posed an existential threat to any effort to preserve the web. Recall that the Task Force on Archiving of Digital Information had understood the significance of copyright years earlier. At the scale that the Internet Archive operated and collected, and with its small staff, negotiating rights on an individual basis would be impossible. The Internet Archive's aggressive collecting of web-based materials put it on uncertain legal ground.

This was especially true as legislative changes had recently transformed the world of American copyright law. In 1978, a revised Copyright Act came into effect in the United States, dramatically expanding copyright's scope. Until then, copyright had to be consciously enacted, but, as copyright scholars Patricia Aufderheide and Peter Jaszi noted, "since 1978 in the United States, all expression that ends up in a fixed medium (and that means everything—your shopping list, the interoffice memo, your kid's homework) is copyrighted by default." Compounding this, these expressions were copyrighted for the period of life of the author plus fifty years after death.[120] This was a big expansion. The previous life of a copyrighted work had, before 1978, been a maximum of two twenty-eight-year terms, explicitly invoked and renewed. To balance this broader scope, the Copyright Act extensively and explicitly codified the practice of Fair Use, which outlined how one could legally use copyrighted material in a variety of domains, from commentary, quoting, scholarly discussion, and satire.[121] In 1998, the Copyright Term Extension Act, also known as the Sonny Bono Copyright Term Extension Act in honor of the late congressman and musician, extended all existing copyright terms by twenty years.

For Kahle and the Internet Archive, the ever-expanding nature of copyright was an existential threat to the vision of the institution that they were seeking to build. Kahle did not mince words in our interview: "it's the stupidest, worst future-destroying, library-destroying thing the United States has done since, I don't know, blowing up Hiroshima or something."[122]

This was especially true as the digital age brought dramatic expansion of the ability to both publish and *re*publish. This had dramatic implications for rights holders. In the world of computing, an open approach to intellectual property was a hallmark of the early field. As developers tinkered and programmed, many brought a looser approach to intellectual property as they built on each other's work. A collaborative spirit of sharing and community in part defined the field.[123] As intellectual property historian Adrian Johns has explained, due to this, "a fundamental fault line [emerged] around digital creativity and intellectual property."[124] Bill Gates, cofounder of a then-young Microsoft, decried widespread software piracy. He saw it as theft, leading him to pen a 1976 open letter that assailed "not just the particular perpetrators of the 'theft' (as he called it), but, in sweeping terms, the culture that endorsed such actions."[125] This was an early glimmer of the tensions between the expanding power of copyright and intellectual property on the one hand, and a digital culture that enabled the dramatic copying and transmission of material without much respect for intellectual property on the other.

The Internet Archive has been entwined with copyright issues since its founding. Indeed, for Kahle copyright, as opposed to technical barriers, represented the "biggest problem of the whole digital dark age."[126] After the Wayback Machine's launch, copyright lawyer and activist Lawrence Lessig predicted to the *New York*

Times that "holders of copyright will eventually drag Mr. Kahle into court." Yet Lessig did wonder, more optimistically, if the Wayback Machine would help illustrate to a broader public the stakes at hand if copyright got too restrictive.[127] The copyright landscape that the Internet Archive was operating in was, indeed, complex. Writing in *Salon,* Katharine Mieszkowski mused on whether the current state of copyright might "leave nothing but junk?":

> The Internet Archive's non-profit status may help it avoid some legal challenges, but it is still not immune from basic copyright concerns. The problems that arise aren't likely to be entirely solved by blocking access to individual sites within the archive. That's because the copyright to the content of any given site doesn't necessarily reside with the operator of that site. For instance, a wire service, such as the Associated Press, might balk when it discovers that thousands of its stories, published on other sites, can be freely visited in the Internet Archive Wayback Machine. The testy members of the National Writers Union may also view the archive as an unauthorized and uncompensated republishing of their work. There's also the tricky question of what happens if a settlement in a lawsuit requires that libelous material be removed from a Web site, yet the original lives on in the archive?[128]

Many of these initial apocalyptic visions would, fortunately, not come to pass. This was primarily due to the Internet Archive's opt-out mechanism.

Since 1996, the Internet Archive has used an opt-out basis for content owners to either ask for their sites not to be crawled or to have their material retroactively removed. To do so, the Internet Archive originally leveraged the robots.txt protocol. Robots.txt, created by Dutch software engineer Martijn Koster in 1994, was a voluntary protocol that allowed a website to request that crawlers or bots should refrain from visiting a given site. Designed in a context of bandwidth scarcity—where many websites had limits on the amount of data that could be transferred out of it before hitting a limit that would lead to a "this site has reached its cap" message to visitors—the robots.txt protocol let web developers preserve bandwidth for actual users.[129] The robots.txt protocol evolved from that limited scope into a broader protocol that would keep a site out of a search index. The early search engine AltaVista, founded in December 1995 and one of the leading search engines over the next half-decade, promoted the use of the robots.txt protocol and the little file thus became one of the "defaults" of the web.[130] Robots.txt appealed for two reasons. First, it was and is a de facto standard that crawlers can check before harvesting a website to ensure they are welcome. Second, as robots.txt needs to be placed in the root directory of a server, anybody who can add or modify a robots.txt file at the very least has high-level access to the web server. Yet these advantages mean that a level of technical access and acumen are needed to use it. Those using a third-party platform such as GeoCities in the 1990s or WordPress today do not usually have the privileges or technical knowledge to implement the robots.txt protocol.

The use of the robots.txt file offered two key advantages for the Internet Archive. Not only would the Internet Archive then refrain from crawling a website that asked robots to not visit the site, but after the Wayback Machine launched in 2001, robots.txt could also retroactively remove sites from being accessed. Yet something beyond this adoption of a de facto standard was needed.

In 2002, the Internet Archive convened "outside archivists, librarians, and attorneys" at the University of California, Berkeley, to discuss (and codify) the thorny issue of collections and opting out.[131] The ensuing "Recommendations for Managing Removal Requests and Preserving Archival Integrity," or "Oakland Archive Policy" became the Internet Archive's guiding policy for handling take-downs and opt-outs.[132] There is some irony that the policy informing the Internet Archive's take-down procedures, rhetorically invoked as an outside authority governing these forms of requests, was and is maintained by the Internet Archive itself. Regardless, the policy brought transparency and consultation to bear on this eternally challenging problem. The Oakland Archive Policy was informed by the American Library Association's Library Bill of Rights, the Society of American Archivists' Code of Ethics, and the International Federation of Library Association's "Internet Manifesto."[133]

The Oakland Archive Policy outlines several main types of removal requests. The first are those made by the content's webmaster ("typically for reasons of privacy, defamation, or embarrassment") but also third-party requests for content removals based on reasons as diverse as the 1998 Digital Millennium Copyright Act, intellectual property claims, controversial content ("e.g. political, religious, and other beliefs"), personal data, and government requests. For personal data, the Oakland Archive Policy also recommended that personal data should be understood as the property of the affected person (i.e., medical information about a person should be removed if the subject requested it rather than deferring to the site itself). In general, while court orders must be respected, the policy holds that "libraries should challenge censorship in the fulfillment of their responsibility to provide information and enlightenment."[134] For individual webmasters, the policy was that:

1. Archivists should provide a "self-service" approach site owners can use to remove their materials based on the use of the robots.txt standard.

2. Requesters may be asked to substantiate their claim of ownership by changing or adding a robots.txt file on their site.
3. This allows archivists to ensure that material will no longer be gathered or made available.
4. These requests will not be made public; however, archivists should retain copies of all removal requests.[135]

Examples were given as part of the policy on how to use the robots.txt file. At the Internet Archive, there was also an email address to contact if one wanted their site to be taken down—essential to those for whom the robots.txt file was inaccessible or not relevant.

While copyright was a key concern for the Internet Archive, the organization encountered more issues as it expanded its vision of what a twenty-first-century universal library would look like. What began as tape drives at a decommissioned army post would lead to advocacy on the steps of the American Supreme Court. As Lessig noted: "Kahle and the Internet Archive suggest what the future of libraries or archives could be. *When* the commercial life of creative property ends, I don't know. But it does. And whenever it does, Kahle and his archive hint at a world where this knowledge, and culture, remains perpetually available."[136] Yet open access to knowledge would not come easily.

Copyright and the Quest for Universal Access to All Knowledge

In 1995, Eric Eldred, a retired computer programmer in New Hampshire, decided to build an online library of public domain books. Eldred scanned them, transcribed them, and put them online.[137] Three years later, in 1998, Eldred was planning to post Robert Frost's 1923 poetry collection *New Hampshire* when the American Congress passed the Sonny Bono Copyright Term Extension Act. Overnight, the copyright term for works *still* in copyright published before 1978 was extended from 75 to 95 years. What that meant is that while something published in 1922 became public domain in 1997 and remained there, anything published in 1923 would remain in copyright until 2018 as opposed to 1998.[138] It was a sudden shift.

Not wanting to wait another two decades, Eldred decided to go ahead and publish the scanned and transcribed version of *New Hampshire*. This put Eldred into dangerous legal territory, perhaps more so than one would think when contemplating the uploading of a 1923 poetry collection. As Lessig explained, "because of a second law passed in 1998, the NET (No Electronic Theft) Act, his act of publishing would make Eldred a felon—whether or not anyone complained."[139]

There would be no need to prove damages. What began as a fight over a collection of Robert Frost's poetry would eventually reach the American Supreme Court.

Eldred partnered with Stanford professor and attorney Lawrence Lessig and several other parties to preemptively challenge the government's copyright extension. American copyright laws stem from Article I, Section 8, Clause 8 of the US Constitution, which holds that Congress has the power to "promote the Progress of Science and useful Arts, by securing for limited Times to Authors and Inventors the exclusive Right to their respective Writings and Discoveries."[140] Lessig and Eldred filed a lawsuit in DC federal district court in January 1999, arguing that the continual extension of copyright terms in legislation such as the Sonny Bono Copyright Term Extension Act violated the Constitution's "limited times" provision.[141] This was because, in theory, Congress could have kept extending copyright by continually passing legislation, effectively making the copyright powers de facto indefinite and not limited at all. Eldred and Lessig also alleged that these extensions were a violation of their First Amendment rights to free speech. Unsuccessful in forcing a hearing by the DC District Court, the two men appealed to the DC Court of Appeals.

While unsuccessful in their arguments before a small panel of the DC appeals court, Lessig and Eldred were emboldened by a dissenting opinion. Judge David Sentelle, a conservative jurist, agreed with their argument, noting that the court should deny Congress's power to interpret the meaning of a "limited time."[142] After another unsuccessful appeal to have the court as a whole hear their case, Lessig and Eldred were further encouraged by another judge—this one a liberal—who agreed with his ideological opposite that Congress was out of line. Making a final appeal to the Supreme Court, in February 2002, the plaintiffs and media alike were surprised to learn that the Supreme Court—which considers only a fraction of cases that are appealed to it—would hear their case in October 2002. The stage was set for a pivotal showdown concerning the American copyright system.

Enter the Internet Archive. Kahle's idea of "Universal Access to All Knowledge" was by then expanding to encompass physical books. Kahle had started with the web, moved into television around 2001, and by 2002 was beginning to digitize books. As Kahle noted, "the Internet is the 'information resource of first resort' for millions of readers . . . I found this exciting and frightening . . . But the Net doesn't have the best we have to offer."[143] For that, books would need to join them. Starting in September 2002, inspired by *Eldred v. Ashcroft* as well as the more general aims at the heart of the Internet Archive, Kahle worked to transform the text files found on Project Gutenberg (a pre-web project to digitize public domain classical works) into print-on-demand books. The Internet Archive would format the texts using a word processor, lay them out, and then print and bind them.

The Internet Bookmobile garnered the most attention. Three weeks or so before the case was to be argued, Kahle and the Internet Archive quickly built a bookmobile to set out across the country.[144] Kahle purchased a 1992 Ford Aerostar minivan, equipped it with a binding machine, a MotoSAT satellite dish, a color printer, and a paper cutter. A sign was hung on the side declaring "1,000,000 books inside (soon)."[145] The idea was compelling. Richard Koman wrote in *Salon* that the "Internet can be a digital library filled with the full array of human knowledge. Technology allows us to bring this massive resource anywhere, not just for reading on screen, but for creating books themselves."[146] The idea was that the Internet Bookmobile would in effect be a "print-on-demand-mobile." A visitor could visit the Bookmobile, select a public domain book, and walk out with a bound volume. What better illustration of the power of public domain?

With *Eldred v. Ashcroft*'s hearing looming, Kahle drove the Bookmobile from San Francisco to the steps of the Supreme Court in Washington. The Bookmobile arrived on 9 October 2002, when the Supreme Court heard the case. Joined by his son, two friends, and the freelance writer Richard Koman, the Bookmobile visited along the way a Salt Lake City school, a librarian conference in Columbus, the Akron International Inventors Museum, Pittsburgh's Carnegie Library, and a school in Baltimore before arriving in Washington.[147] It was a vivid demonstration of a "library truly free to the people."[148]

En route, the Bookmobile made a surprise stop in Urbana, Illinois, to visit Michael Hart, founder of Project Gutenberg. Hart had started Project Gutenberg in 1971 by manually typing works of literature and initially sharing them on the ARPANET, then continuing this work on the internet and the web. By 2002, Project Gutenberg had enlisted a hundred volunteers to collectively digitize 6,000 books. Kahle tried to convince Hart to take the Internet Bookmobile on another cross country tour the following year, and he assured Hart that he would help him get to 10,000 digitized books. With some (but not all) tongue in cheek, Hart explained to Kahle and Koman the real goal behind digitizing all these books. "You know that episode of 'Star Trek,' when they look in the computer to find some 20th century book that tells them what to expect when they go back in time," Hart explained. "How do you think those books got in the computer? That's me."[149]

The last stop before the Supreme Court was the main branch of the Carnegie Library of Pittsburgh. The classical building, dating from 1895, has the words "FREE TO THE PEOPLE" inscribed in its imposing stone façade. What better motto for both the Bookmobile's efforts, as well as those of Eldred and Lessig. Speaking to the media beneath the inscription, Kahle argued that he had shown that it was "possible to create million-book libraries for free," and that when it came to *Eldred v. Ashcroft*, "the public domain is on trial."[150] As Koman noted, "FREE TO THE PEOPLE" fundamentally captured Kahle's vision: "A massive library containing the full breadth of human knowledge and experience, freely and easily accessible to everyone on the planet."[151] The image in figure 3.1 shows the Bookmobile in action beneath the inscription.

The Bookmobile's next and final stop was the steps of the Supreme Court in Washington, DC, while Eldred and Lessig argued their case inside. *Eldred v. Ashcroft* was decided in a 7–2 ruling against Eldred. The majority opinion, written by Justice Ruth Bader Ginsburg, noted that as long as Congress set time limits for extensions (which, to her, satisfied the "limited times" requirement) then continued extensions were constitutional. In hindsight, Lessig felt he had erred in not stressing the harms caused by limiting the public domain. Lessig had focused on structural and legal arguments rather than impact.[152] The public domain remained limited, and indeed, it was only in 2019 that the public domain again expanded in the United States. Only then would Robert Frost's *New Hampshire* be in the public domain.

In late 2004, the Internet Archive began digitizing books at scale, representing a dramatic expansion of its scope. Rather than building the web's memory, this represented a shift to making memory *on* the web. In December 2004, the Internet Archive began to work with libraries around the world to incorporate public domain, scanned works into their collections. A month later, the Internet Archive had around 27,000 books online.[153] At the same time, in December 2004, Google let its own consortium challenge publishers with its "Library Project," alongside several academic libraries. The Authors Guild would later file a lawsuit accusing Google and partners of "massive copyright infringement."[154] Meanwhile, the Internet Archive's smaller, public domain project—now known as the Million Book Project—proceeded. The Internet Archive would initially collaborate with Indian and Chinese libraries.[155]

In October 2005, the Internet Archive helped to launch another large-scale book digitization project: the Open Content Alliance, or OCA.[156] Kahle announced it with rhetorical flourish: "The opportunity before all of us is living up to the dream of the Library of Alexandria and then taking it a step further—Universal access to all knowledge."[157] If Google Books was limited to snippet views, the OCA would focus on the public domain to provide books in their entirety. "We don't talk about snippets. We talk about books," declared Kahle.[158] Microsoft joined

Figure 3.1 The Internet Archive Bookmobile in front of the Carnegie Library of Pittsburgh main branch; Brewster Kahle (*right*) and his son, Caslon (*left*).
From Internet Archive, "Bookmobile Photos," https://archive.org/texts/bookmobile-photos.php

the OCA as well, setting up a showdown between tech titans Google and Microsoft over digitizing books. It was not an inevitable clash—Kahle had greatly hoped that Google would join the OCA as well—but the media could not avoid a Google versus Microsoft framing.[159]

Through 2006 and 2007, then, the Internet Archive was part of the "moonshot" era to digitize *all* printed human knowledge. This was explicitly invoked by Kahle as he spoke at the OCA's kickoff: "It will be remembered as one of the great things that humans have ever done, up there with the Library of Alexandria, Gutenberg Press, and the man on the moon."[160] The excitement was palpable. "This is our chance to one-up the Greeks!" Kahle exclaimed to a *New York Times* reporter, "It is really possible with the technology of today, not tomorrow. We can provide all the works of humankind to all the people of the world. It will be an achievement remembered for all time, like putting a man on the moon."[161] The reason behind this made sense as humanity entered the age of search. People could search for almost anything using search engines, but books remained inaccessible.

By 2008, the Internet Archive declared its intent to "archive" "a single scanned copy of every book ever published."[162] It would do so through global library partnerships. The final stage of the Internet Archive's quest for universal access to knowledge happened with its Physical Archive in 2011. Purchasing a warehouse near San Francisco, Kahle spent $3 million with another bold goal. "We want to collect one copy of every book," he declared to the *New York Times*.[163] Books, as well as movies, records, and beyond, would be housed in shipping containers, both to preserve culture but also out of a realization that if the Archive held a physical version it could "share the digital one."[164]

All of this represented a dramatic expansion for the Internet Archive. In a few years, the library had evolved from providing access to archived websites to archived content of all kinds: to books, software, government documents, appliance manuals, and beyond. In doing so, the Internet Archive challenged ideas of copyright, the traditional organization of knowledge, and crucially, made a case for universal access to all knowledge.

Conclusion: A Sustainable Institution?

From Kahle's first effort to make "everyone a publisher" with WAIS to the early stirrings of a side project to create a backup of the still-new web, the Internet Archive grew into the internet's universal library. Emerging from a milieu that was beginning to value digital preservation, the Internet Archive sprang into action in a context of a receptive public, commentariat, and academic community. Primed by the cultural conversations that convinced many of both the value of digital preservation as well as the specter of a digital dark age, the Internet Archive was positioned to thrive. While much of this chapter's story has involved Kahle, who provided not only capital but also the intellectual energy behind the project, the organization's longevity speaks to the broader community role that it filled.

Sustainability had long been a concern about the Internet Archive: could one trust a singular organization with the long-term stewardship of the internet's memory? If the Internet Archive failed, would we not simply again have a digital dark age—compounded by an increasingly single point of failure? Were the risks too much? In 2003, Roy Rosenzweig had worried about this. "Even more troubling, it has no plan for how it will sustain itself into the future," Rosenzweig noted. "Will Kahle continue to fund it indefinitely? What if Amazon and Alexa no longer find it worthwhile to gather the data, especially since acquisition costs are doubling every year?"[165] Buried in a footnote to that statement was some encouragement: "Insiders have commented to me that the [Internet Archive] would disappear if Kahle left the project. But there are very recent signs that the [Internet Archive] is broadening its base of financial support."[166] These worries persist into today. Richard Ovenden noted that sustainability was what concerned him most about the Internet Archive and its "modest funding base." In his mind, eventually "it must become part of, or allied to, a larger institution, one which shares its long-term goals to preserve the world's knowledge and make it available."[167]

The Internet Archive is more sustainable than these observers give it credit for, however. Many of the partnerships with national libraries, and later universities and other cultural organizations, discussed in passing in this chapter and at length in chapter 4, would indeed help with the Internet Archive's sustainability. Public filings with the American Internal Revenue Service, in the form of Form 990 filings which provide financial information on nonprofit organizations, illustrate this, although the complexity of financial data means that they do need to be used with caution. The Library of Congress first provided sums ranging from $65,000 in 2001 to $787,974 in 2004 (which would include both web archiving as well as perhaps some other digitization projects), against the Internet Archive's overall budget of $1.5 million or $4.5 million, respectively. By 2006 "crawling services"—or "fees collected for providing web crawling and hosting services"—amounted to $5,362,188 income versus an overall budget of $9,419,040. By 2008, this was $10,088,116 against total overall costs of around over $11 million.[168] Given the wide variety of other activities the Internet Archive is involved in, from book scanning to digitization and beyond, the web archiving operation appears to be sustainable in and of itself.

Figure 3.2 The Internet Archive in San Francisco. Photo by the author

Furthermore, the Internet Archive's independence and agility have allowed it to tread a path not open to larger, national institutions. It is thus embedded as part of a larger global library community. Partnerships take shape in scanning centers, operated around the world to help institutions digitize their print heritage, as well as their "Archive-It" subscription service, which launched in 2006. Institutions partner with the Internet Archive to select websites that should be archived. Whether done for institutional compliance, such as backing up websites, or in the service of history, this model today sees over 94% of North American institutions who web archive turn to Archive-It as a service.[169] Modern web archiving has been shaped by the Internet Archive.

The Internet Archive is also a product of the internet itself. Throughout this chapter, we have seen how it has been profoundly influenced by the internet's ever-evolving culture. The Internet Archive, in its association with the struggle for copyright reform as well as its emphasis on free access, came to articulate a vision of culture as free in the same way that early free software pioneers understood software to be *free* in the sense of a fundamental right to the freedom of speech (as opposed to, as they put it, free as in "free beer").

By 2009, the increasingly-not-so-small operation in the former post office at the Presidio had begun to outgrow its small building. By then, the Internet Archive had the largest and most comprehensive web archive, led a large global book-scanning operation, and was providing global leadership to digital preservation efforts across the world, as we will see in the next chapter. An opportunity reared its head in 2009, when the Fourth Church of Christ Scientist went on sale in San Francisco's Inner Richmond district—23,000 feet of neoclassical architecture, which happened to resemble the Internet Archive's classical logo. The building (figure 3.2) is a fascinating place, even more so in a city with stratospheric real estate costs that would on the surface seem to preclude hosting physical servers a mere twenty-minute drive from downtown. As Venkat Srinivasan wrote in *Scientific American*, "there is no pressing need to house a bunch of servers in a church, in a residential neighborhood, and especially in a city where it is increasingly difficult to survive on a librarian's salary. No need, that is, except the possibility of being a community fixture."[170]

This last point is key. The Internet Archive's new cathedral-like home heralded ever-increasing amounts of community engagement from the organization throughout the Bay Area and the world. Responding to the 2008 financial crisis, in 2011 Kahle and the Internet Archive opened up the Internet Archive Federal Credit Union in a low-income community near New Brunswick, New Jersey (owing in part to regulatory burden, the Credit Union closed in 2015).[171] In 2015, near the Internet Archive's building, Kahle set up a nonprofit corporation to develop "Foundation Housing." Given San Francisco's high cost of living, the organization purchased apartment buildings and then cheaply rented them out to nonprofit employees.[172]

Between 1996 and today, the Internet Archive emerged as the internet's library. While it was a pioneer, it was also a response to larger intellectual forces. We can see this borne out in the next chapter when we turn to other early web archiving operations. Emerging at the same time, they would take different approaches to saving culture.

CHAPTER FOUR

From Selective to Comprehensive

National Libraries and Early Web Preservation

"The alternative to progress . . . is simple: the Library of Congress could become a book museum," noted the authors of *LC21: A Digital Strategy for the Library of Congress*, a report commissioned and released in 2000 to help guide America's de facto national library into the digital age. Being a book museum would not be a good thing. For "a library is not a book museum. A library's value lies in its vitality, in the way its collections grow, and in the way that growth is rewarded by the diverse and innovative uses to which its collections are put."[1] This aptly summarized the challenges facing libraries in the digital age.

Libraries—research, public, and national alike—were grappling with the challenge of moving from a model based on physical objects to one based on networked information. A considerable challenge was the ever-present problem of copyright. Much of the borrowing and access rights for national libraries, even when material was collected via legal deposit, was predicated on possessing physical objects. If the "original" copy of a publication sat on a corporate or personal server, it would be a reproduction when copied for library preservation.[2] For national libraries, stewards of national heritage and essential components of a nation's historical memory, the stakes were high as they adapted to the digital age.

This chapter explores the evolution of web archiving at national libraries at four pivotal early institutions: the national libraries of Canada, Sweden, Australia, and the American Library of Congress. These institutions were chosen for their pathbreaking role in national web archiving. Yet it is worth noting, as will become clear, that despite the expansive visions that a "national library" might conjure, these were small operations with modest funding. They generally received no specific financial support and thus needed to be funded by existing operational funds. However, this required sufficient operational funds and confidence in their ability to deliver on the core mandate while exploring these questions, meaning that only a small handful of affluent institutions could do so. The visions that they shaped would build the foundation for today's widespread web preservation. Ultimately, the end goal of comprehensive legal deposit of websites, which would bring them on par with books and recognize their cultural value, would be informed by different approaches. The Canadians, Australians, and Swedes adopted different approaches to collection. Some crawled broadly, others selectively.

These institutions did not develop their programs independently of each other. Terry Kuny, at the National Library of Canada, recalled active mailing lists. These helped build an "invisible college to bring people of like minds together."[3] A conversation around the future shape of web archiving and born-digital preservation occurred. In 1998, national libraries in Sweden, Australia, and the Netherlands established PREWEB: Preserving the WorldWideWeb, "an international discussion list and link collection," helping to form that invisible college.[4] Links between libraries were also made possible by the web itself. The web, as Margaret Hedstrom recalled, "made connections that just weren't feasible [before] . . . digital preservation became an international phenomenon, in part because of what the world wide web enabled."[5] By 2001, practitioners began to gather for the International Web Archiving Workshops convened by Julien Masanès.[6]

This chapter explores the various ways that national libraries approached the defining question of their web archives: would they be selective and actively curate their web archives, or would they try to be comprehensive and collect everything? Today, the question has been largely settled in favor of comprehensiveness. While many national libraries continue to be selective, it is now usually framed as an intermediate step on the way to comprehensive crawling. This was an unsettled question over the first fifteen years of web archiving, however. Some libraries worried that the mass collection of information would lead to large, low-quality, and ultimately less useful, collections. If the Internet Archive had led the way on mass collecting, perhaps a more selective approach would better steward cultural heritage? As they confronted digital abundance, national libraries would adopt different approaches.

Selection versus Comprehension: The (False) National Library Dilemma

Today the debate between selection versus comprehension seems like a false binary. To many in the 1990s, however, it appeared to be a stark choice. Julien Masanès explains: "A recurrent theme in the literature on Web archiving is somehow simplistic opposition between manual selection and bulk automatic harvesting allegedly considered as unselective. The former is misleadingly supposed to be purely manual whereas the latter is similarly falsely considered as comprehensive. We prefer to insist on the fact that Web archiving always implies some form of selectivity, even when it is done at large scale and using automatic tools."[7] In other words, even the most comprehensive crawl requires active scoping. Questions such as where to start, how deep to crawl, how long until the crawlers should collect without stopping, are all curatorial decisions. Algorithms are designed and controlled by people. Furthermore, true comprehensiveness is impossible given the web's ever-changing nature. Comprehensiveness thus refers to objective and aim, as opposed to a mythical state of all-complete and all-finding crawling.

The main argument behind being as exhaustive or comprehensive as possible is that historians do not know what might be useful in the future.[8] There are many reasons to collect as much as possible. These range from biographic (capturing the homepage of a future celebrity, activist, or politician who is then unknown), cultural or social (advertisements, trends, and subcultures), and basic logistics (easier and even potentially cheaper to collect it all since it can be better automated).

There will almost certainly be future lines of historical inquiry that we can scarcely imagine today. This reasoning informed the Internet Archive, which attempted to capture as much as possible and worry about implications later. Among national libraries, the National Library of Sweden would emerge as one of the leading voices behind this "comprehensive approach." Most national libraries would be more circumscribed, whether due to resource constraints, legal issues, or collection philosophy.

Given constraints, the temptation was to be selective. This was due both to size and technical issues, as well as the need to have firm scoping criteria that would delineate a collection. Scoping would keep a web archive at a manageable size, allowing for cataloging. Archived websites could then be treated more akin to traditional "publications." Similarly, in the absence of legal frameworks, a website-by-website permissions process was possible only if approached selectively. With broad web crawling, the scale of collecting means that permissions become effectively impossible.

The selective model was pioneered in 1994 with the National Library of Canada's Electronic Publications Pilot Project. The EPPP considered hyperlinks, the expansive and sprawling nature of publications in the digital age, and it recommended that collection efforts needed to be targeted and selective for the ensuing collection to be usable. After this pilot, this approach would inspire the leading selective archive: the National Library of Australia's PANDORA project. In other words, the selectivity versus comprehensiveness discussion—exemplified by the

Canadians and Australians on the one side versus the Swedish and the Internet Archive on the other—becomes a useful way to introduce and frame their projects. Yet, in many ways, this framing admittedly involves a bit of authorial convenience on my behalf. For these approaches were also informed by the scarcity of resources as well as legal constraints. In other words, a selective approach was often the only realistic one, given available options.

The first two initiatives in the world to tackle born-digital publications occurred at the Netherlands' Koninklijke Bibliotheek and at Canada's National Library.[9] The Koninklijke Bibliotheek example is interesting, as its library has existed without legal deposit since its 1798 founding. The "Dutch Depository of Electronic Publications" project aimed to gather "all electronic publications produced in The Netherlands," including full-text documents, databases, and multimedia files.[10] The Netherlands' pilot ran through late 1995, with extensive tests of its processing workflows. It focused on electronic journals, educational software, and "other" publications including HTML documents. Yet the Canadian example became inspirational to web archival projects that would follow, making it a natural starting point for this chapter. Beginning in 1994, a small group of librarians and digital preservation experts began a pilot project that would be the first national library effort to preserve material on the then-new web.

First Mover: The National Library of Canada's Electronic Publications Pilot Project

The National Library of Canada inspired subsequent web archiving projects. Its early track record built a foundation for the projects that followed. The National Library of Canada (NLC), established in 1953, had for decades been a leader in information policy and digital libraries.[11] This was partially due to the amount of information that had been flowing across the American-Canadian border since at least the 1970s. NLC librarians were extensively engaged in international conversations with their American colleagues, requiring considerable thought about standards and interoperability. Partially in recognition of this unique context, the International Federation of Library Associations and Institutes established its office for its Universal Dataflow and Telecommunications office at the NLC, with its mandate for the "international and national exchange of electronic data."[12]

The nearby National Archives of Canada (Public Archives of Canada until 1987) was also a leader in the field. As Hedstrom explained, "a lot of the [international] leadership for that [electronics records] work was coming out of the Public Archives of Canada."[13] In December 1992, the National Summit on Information Policy was held, which brought together libraries, governments, and information professionals in an attempt to reconcile disparate information policies across the country.[14] As this was a relatively decentralized federation, competing provincial standards and approaches across Canada presented challenges. But, by 1994, the Canadian government was already moving to ensure that all government publications would be digitally available.[15] As a relatively early adopter, with a national library engaged in electronic records, the stage was set for an innovative program when it came to new media.

The NLC's digital leadership was also enabled by an expansive mandate stemming from its founding legislation. The NLC argued that its mandate to collect information applied "to all Canadian published information, wherever it is located and in whatever format."[16] While the NLC was initially given a formal mandate only to collect "books," as early as 1952 the National Library of Canada Act crafters noted that "book" should be understood as inclusive of "library matter of every kind, nature, and description."[17] In 1969, the act was amended and made explicitly inclusive of "any document, paper, record, tape or other thing published by a publisher, on or in which information is written, recorded, stored, or reproduced."[18] Accordingly, in 1969 the library began acquiring sound, in 1988 microforms, and in 1993 videos and CD-ROMs.[19]

This was the organizational context in 1994, when the NLC looked at how to preserve documents published on the web. Librarians saw that the publishing world was being reshaped by newly emerging communications platforms. Scholarly journals were being shared via Gopher, WAIS, and the web, as were novel forms of electronic books (hypertext!) and other publications.

Between June 1994 and summer 1995, to explore what born-digital acquisitions would look like, the NLC launched the Electronic Publications Pilot Project (EPPP).[20] The "electronic" part of electronic publications was broadly defined, including "those that are distributed in multiple copies on physical media" as well as "[t]hose that reside on a host computer and are accessible over a communications network." If "electronic" was defined broadly, publications were instead defined narrowly. For the EPPP, publications would refer only to those works that underwent "the same formal preparation activities associated with traditional print publications."[21]

In 1994, these were critical issues. As Kuny, who was then working at the NLC recalled of the EPPP: "It exposed some of the challenges that we had with the fundamentals of digitally preserving things: How do you take in things in very different formats? How do you preserve an early Web object, for example, when you didn't have a container format for a Web object? How do you maintain all the relationships for something on the web? How do you keep a PDF file when the PDF format itself was evolving and changing over time?"[22] Some of this was technical (file formats and storage) and some of this was organizational (in what catalog fields do you record an operating system or file type?).

The EPPP was an opportunity for the NLC to enter the networked digital age. "As the federal cultural institution responsible for Canada's printed heritage," Marianne Scott, then National Librarian of Canada, explained, "the National Library must find new ways of fulfilling its heritage mandate in the information age."[23] The project had six goals: understanding the issues, developing knowledge around online documents, determining broader policies, building the NLC's government information role, providing input for resource requirements, and gaining general institutional experience with internet publishing. It is worth highlighting Scott's role herself. Kuny described Scott as "all the values of librarianship . . . rolled up in one," and an organizational leader who took a keen interest in these new emerging mediums.[24]

The EPPP identified many of the critical issues that would anchor the web archiving conversation over the coming decades. For the Canadians, selectivity was seen as the pragmatic choice. Real resource constraints circumscribed the project from its outset. Even by 1994, the web's scale and the number of electronic publications outstripped the available technical resources. Project staffing resources were generous, as many individuals worked part-time on the project from various departments.[25] The EPPP also fostered initial conversations around how to crawl web-based materials. The project's final report appeared in June 1996, representing the pilot's main contribution to both the NLC and the broader nascent web archiving community.

Topics of discussion in the final EPPP report included copyright, especially conflict between copyright and the need to migrate file formats to ensure long-term preservation. The report also explored versioning (which document is "canonical" when it is constantly changing and evolving), legal deposit considerations, as well as how to catalog documents. As Kuny recalls, "EPPP was a technology demonstrator, and it . . . served an institutional political objective . . . to try and generate the necessary resources to pursue this agenda in the future."[26]

The EPPP argued that the term "publication" needed to be narrowly defined. This decision would have the most impact on national libraries inspired by the EPPP. Even for the web, the "characteristics of traditional publications" were emphasized. The report stressed an acquisition approach centered around online periodicals and scholarly journals, as opposed to personal homepages or what would later be classified as "ephemera." The pilot conceded the difficulty and subjectiveness at play to define what a "publication" was versus "ephemera." Indeed, in the formal report, these criteria were literally described as an "I'll-know-it-when-I-see-it" statement.[27] Secondarily, given the narrow definition of a publication, a preserved publication would need to be removed from a broader hyperlinked web. This was an important distinction.

In the case of a web-based hypertext publication, where links were integral to the work itself, the report was candid about the difficulties curators faced. Most web documents contained links, and those links (and the ability to interact with them) were part of the document. To maintain a publication's integrity, should everything that the document linked out to be preserved too? The EPPP report, written in June 1996, grappled with this at length:

> What, then are the options? One could, copyright permitting, bring in the files associated with every hypertext link in a publication and store them with the files for the publication themselves. Or, one could choose to pull in only a portion of the links, such as those to Canadian sites or to Canadian publications elsewhere. One could also choose to regard a hypertext publication as analogous to a TV program guide, which points to program items that may or may not be archived somewhere, but may not be viewed beyond the limited lifespan of an issue of the guide. Despite the leads being blind, the "pointing tool" retains a certain social and historical value.[28]

The latter point was analogous to the dire access situation of television archives, where historians, indeed, rely on the catalogs rather than the recordings themselves.

Ultimately, the EPPP pilot recommended that "only those links which resided on the same Internet domain as the original document should be brought in and placed on the NLC server."[29] The line between publication and broadcast was also drawn. Up until that point, the report noted, "it has been NLC netsurfer experience that the non-directory type of Web site, consisting mainly of links, is ephemeral and more analogous to a broadcast than to a publication."[30] This notably set a tone within the institution that would lead toward a recommendation to be *selective* rather than comprehensive. The final report was clear: "Electronic publications should be acquired on a selective basis."[31]

As for web archiving, the EPPP would be a high point as NLC priorities shifted. The NLC continued to harvest selected publications and government webpages but would not expand this limited scope until 2005. Much of this was the reality of a time-limited short-term project coming to an end without a real successor or long-term strategy in place. It "just kind of disappear[ed] over time, and if there is no will at all, it is done," explained Kuny, summing it up as: "no responsible authority, no commitment to migration over time, no financial resources."[32]

The EPPP's legacy would later be felt in the expansiveness of the enabling legislation for Canada's newly amalgamated national memory institution. This new institution, Library and Archives Canada, was a merger of the NLC and the National Archives of Canada. The 2004 Library and Archives Canada Act gave the institution the power to take a "representative sample of the documentary material of interest to Canada that is accessible to the public without restriction through the internet or any similar medium." Another section noted that the power of legal deposit extended to "publications that use a medium other than paper."[33] This was forward thinking for 2004.

The laws, however, would not be used to their fullest. Library and Archives Canada's Web Archiving Program was founded in 2005. While initial plans explored whether it could collect the entire Canadian .ca domain, it was soon downscaled to a selective approach "when the scale and technical complexity of this endeavour became clearer."[34] Thematic collections were actively curated from 2009 onwards, complementing government data harvests, leading to a small but significant curated collection of Canadian material. As of 2019, Library and Archives Canada had approximately 50TB of web archival collections, slightly more than half of which was data from federal government websites.[35] The institution continues to selectively capture Canadian web content, but it is a relatively modest program.

Yet for a period between 1994 and 1996, national libraries around the world looked to Canada as they explored these novel questions. The EPPP set the stage for the conversations that would unfold, particularly in Australia and later the United States, setting the stage for a selective web archiving ethos. But before we pivot to Australia to see how this intellectual current continued, let us turn to Sweden and national library comprehensive collecting.

Kulturarw3: The Advent of Comprehensive National Library Collecting

The National Library of Sweden, or Kungliga biblioteket (KB), adopted a different approach than the Canadians. Whereas the EPPP pioneered a selective approach, the KB elected to exhaustively collect the Swedish web. In September 1996, the Kulturarw3 project began.

Whereas the EPPP more narrowly focused on electronic journals, Kulturarw3 had a broad mandate from its inception. Its goal was to capture Swedish culture as expressed online. The project's name reflected this. As project leader Johan

Mannerheim explained: "The name Kulturarw3 means Cultural Heritage in Swedish but is properly spelled with a 'v' at the end. The 'w' has the same sound value in our language and we have indexed it to point out that the WWW or World Wide Web not only is something new and modern, but also part of our cultural heritage."[36] From its inception, Kulturarw3 was framed as adopting a different approach than Canada had. The National Library of Sweden adopted a comprehensive approach for two reasons. First, it worried about the difficulty of proper selection and worried that future researchers would criticize the choices that it made.[37] Second, and provocatively, the Kulturarw3 team took the position that the labor costs of selective curation would outweigh the technological costs of storage. The hopes were that a comprehensive approach would achieve three goals: gather the most material, be less labor intensive, and leverage the ever-falling price of data.[38] The relatively small population of Sweden—just shy of nine million people in 1996—also helped.

The Kulturarw3 team met for the first time in August 1996.[39] The emphasis was on building the KB's permanent collection. Beyond a few initial thoughts that perhaps they would eventually adopt a one- or two-year waiting period and somehow provide remote access, the discussion was deferred. The Kulturarw3 team looked ahead to the scale of the problem before them. They predicted that as a new generation, accustomed to using the internet, entered their working ages there would be an explosion of creative web use, including audio, video, and other applications.[40] With this looming in the not-so-distant future, it made sense to begin collecting as soon as possible, rather than waiting for final decisions on access.

Kulturarw3 launched in September 1996 after successfully hiring a project manager. Reflecting my earlier cautionary notes around drawing too firm a boundary between selective and comprehensive approaches, while the public discourse around the project noted how expansive the Swedish approach was when compared to other libraries, in reality Kulturarw3 consisted of a mixture of annual, automated crawls as well as selected pages that would be more frequently collected. Selectivity complemented comprehension.

Even with the relatively small size of the Swedish web, the team confronted an almost overwhelming amount of data. The team estimated that there were approximately 700,000 to 1,000,000 Swedish pages, averaging 20,000 characters each.[41] The first crawl began in March 1997. Between 24 March and 26 August, the project collected 143GB from almost six million files, storing it to tape. Testament to the Kulturarw3's ambition, a follow-up crawl was launched almost immediately thereafter on 1 September. This second crawl ran until January 1998 and collected almost 10 million pages and 200GB of files, illustrating the growth of the Swedish web in just one year.[42] Crawls grew in scope over the coming years and the scope of the collection criteria also expanded. The first crawl focused solely on the .se domain, but the Kulturarw3 team realized that much of the Swedish web domain was found under other domains, which required collection from domains such as .com and .org as well as .nu (which means "now" in Swedish).[43] By contacting domain registrars and collecting lists of sites with Swedish postal addresses or telephone numbers, a comprehensive collection took shape.

Kulturarw3 engaged with international peers throughout 1997. Its comprehensive approach was novel within the broader national library community. In April, four Kulturarw3 team members visited the Internet Archive, discussing potential opportunities for crawling collaboration and sharing URLs.[44] This was a fruitful meeting. The Swedish visitors noted that the Internet Archive and Kulturarw3 were the sole projects undertaking automatic crawling of large parts of the web.[45] Mannerheim, the project lead, also visited the National Library of Canada and learned that it had no equivalent comprehensive web archiving projects, noting that it was mostly focusing on government publications and a handful of magazines that did not pose copyright issues.[46] When Kulturarw3 was presented at the 1997 International Federation of Librarian Associations conference, the presenters reported back to the KB that the audience was very interested in Kulturarw3, as no other national library had a comparable program in terms of comprehensive coverage.[47]

Sweden's legal landscape led to interesting discussions around access. First, Kulturarw3 would remain a private, dark archive for six years. Access was a difficult issue, bringing together technical problems (tape drive retrieval is slow) with ethical and legal ones. Several legal concepts collided in Sweden with respect to access. Pertinent concepts included legal deposit, freedom of the press, and personal privacy. All three were identified by the project staff as contributing to a unique Swedish landscape.[48] Legal deposit, in force in Sweden since 1661 but expanded in 1993 to include fixed-form electronic documents, underpinned web archiving in Sweden. Second, there was the tryckfrihetsförordningen ordinance. Loosely translated as "freedom of the press," this ordinance essentially gives freedom of the press to all Swedish citizens as well as codified public access to government information. A stated goal behind tryckfrihetsförordningen was to ensure the free exchange of opinions.[49] Third, there was the personuppgiftslagen, or the Swedish Data Protection Agency. This was an agency that aims "to protect the individual's privacy in the information society . . . the DPA ensures that new laws and ordinances protect personal data in an adequate manner."[50] Naturally, copyright was also an ever-present consideration.

Throughout Kulturarw3's early period, the leadership team faced the interaction between these concepts, institutions, and its growing web archive. In 1997, Kulturarw3 staff inquired if tryckfrihetsförordningen would be adapted to include electronic publications, and ultimately realized that they would need to connect with the personuppgiftslagen to craft a pathway forward if Kulturarw3 was to evolve beyond a dark archive.[51]

The government commissioned the Ministry of Education to explore how web archive access could be provided.[52] Their 1998 final report recommended that the KB as well as the National Archive of Recorded Sound and Moving Images assume joint responsibility of preservation and access. Crucially, the report recommended limiting access to "researchers from established research organisations."[53] As Mannerheim editorialized, in his view these limitations "would be contrary to the democratic aim of the Swedish [legal] deposit law to guarantee free access to information."[54] If it was just formal researchers, what about everyday citizens? The Ministry of Education also recommended that the KB accelerate its collecting: four comprehensive sweeps a year, selective curation of material to be collected via more regular active selection, as well as the inclusion of databases, software, and video games.[55] Dialogue continued between the Kulturarw3 team and the government, with web archivists stressing the importance of pointing out the possibilities of web archives to researchers, and crucially, noting that they needed to ensure that future generations could enjoy their right to access.[56]

Exterior pressure eventually cleared the path for access. In 2001, a Swedish citizen complained that the KB had violated the personuppgiftslagen as well as copyright law when collecting online material. An initial ruling in the citizen's favor by the data inspectorate was appealed and the state thus had to legislate to make such collecting compliant.[57] On 8 May 2002, the government announced a new ordinance specifically dedicated to the KB's right to collect websites and provide access. The KB introduced a new policy that would allow for on-site web archive access.[58] A year later, on 16 June 2003, two specialized terminals were situated in the reading room. Each provided access to users who could make specific URL queries. As the Kulturarw3 collection resided on tape, retrieval times were long (several minutes for each page) and, crucially, users could consult but *not* copy consulted material. The specter of a researcher taking handwritten notes or typing on a laptop to record their web archive exploration appeared. This was a model that would later be adopted by other national libraries, such as the British Library and the Bibliothèque nationale de France.[59]

Ultimately, the decision to move ahead with collecting and to figure out access later was prescient. Had the KB waited until an access regime was confirmed, the first six or seven years of the Swedish web's history would have been lost, except for whatever the Internet Archive had preserved. Dark archives are never ideal—preservation without access is hardly preservation—but as an interim approach, it worked well.

The second critical part of Kulturarw3 was the project's emphasis on international collaboration. The national libraries of the Nordic countries (Norway, Sweden, Denmark, Finland, and Iceland) had long collaborated. More recently, they had been exploring the specific problem of how best to share digital resources. This took shape in the form of the Nordic Web Archive (NWA). The NWA's first meeting was held on 19 August 1997, when representatives from the Finnish, Danish, and Swedish national libraries gathered at the KB. Finland had just begun the region's second web archive, beginning its comprehensive crawling equivalent of Kulturarw3 in summer 1997. The NWA would discuss protocols, archival file formats, and harvester requirements with an eye to producing common standards and declarations.[60] Over the coming two years, the NWA devoted its attention to developing a new crawler. The Danish soon became more active in 1998 as they began their own web archive.[61] By 1999, a dedicated NWA employee was hired.[62] In 2003, the NWA released the fruits of its collaborative labor: the NWA Toolset, "a freely available solution for searching and navigating archived web document collections."[63] And, as noted, developers from Iceland and Norway would later collaborate with the Internet Archive on the development of the Heritrix crawler. Such collaboration was key to capture material that crossed national borders. While clearly necessary in a relatively tight-knit region such as that of the Nordic countries, these same principles could be applied to Europe and, from there, the world. By 2002, Kulturarw3 members were in discussions with the British Library and the Bibliothèque nationale de France, sharing experiences as the British and French considered comprehensive crawling of their own web domains.[64]

This last point suggests the ultimate success of the comprehensive national library model pioneered by the KB as early as 1996. Based on *both* improving the historical record and cheaper curatorial costs, this model soon spread to many other European national libraries. That this model was present from the inception of its web archiving program makes the Swedish experiment all the more significant. It also conceptually emerged at the same time as the Internet Archive, complicating the chronology of early web preservation. It was not the only model, however. There was also a good case to be made for selectivity. Stemming from the Canadian pilot, this approach would see substantial realization in the other early pioneering national library project of 1996: the National Library of Australia's PANDORA project.

PANDORA: The National Library of Australia

The PANDORA project, an initialism for Preserving and Accessing Networked Documentary Resources of Australia, emerged around the same time as Kulturarw3. Its institutional inspiration can be traced in part to the EPPP. "The NLA's PANDORA project was among the earliest of the world's web archiving projects," noted PANDORA manager Wendy Smith in a retrospective piece on the Australian web archive. To her, PANDORA began as "a proof-of-concept archive in 1996, based on the intellectual framework of Library and Archives Canada's selective web archiving project."[65] This approach to selectivity would hold for PANDORA's next decade. Yet, whereas the Canadians had stuck closely to e-journals and serials, the Australians would soon expand their scope.

Australia emerged as a web archiving leader, building on the Canadian EPPP.[66] At the 1999 Digital Library Forum in Canberra, while Australian libraries in general looked overseas toward the American Library of Congress for inspiration, one observer noted that "one area where Australia is well ahead of the Library of Congress is in archiving born[-]digital materials such as Webzines, through the NLAPandora Archive."[67] Australia's web archiving leadership was notable despite two major and recurrent shortcomings: the lack of legal deposit power as well as the continual lack of dedicated funding to the archive. As web archiving managers Margaret Phillips and Paul Koerbin noted in 2007, the web archive itself received no specific funding, meaning that funds had to be redirected from core library operations.[68]

The roots of web archiving at the NLA stretch back to discussions around electronic publications and networked information. As Kieran Hegarty notes, collections development conversations began in 1990 and had turned to acquisition by 1993 and 1994. Access in this period, however, took the shape of directing patrons to the live copy or printing an electronic publication for traditional preservation.[69] A better way forward was needed by 1996: the PANDORA project.

PANDORA referred to Pandora's Box. Koerbin recalled that "I assume it was considered neat at the time to reference Pandora's Box . . . that did become tedious."[70] A staff competition was held, "Preservation and Access" led to "PA" and staff then figured out the meaning of the next five letters by November 1996.[71] The idea, however, stretched back to 1994:

> In 1994, when the World Wide Web was just breaking into the consciousness of most Australians, the librarians at the National Library in Canberra were already wondering how to net the shoals of amorphous information floating around in this new medium.
>
> The result is the Pandora project, an ambitious electronic archive that will preserve home pages and online documents judged of national significance for "hundreds of years" in library databases.[72]

The archive was born out of the broader fears around cultural loss and a digital dark age. Julie Whiting, PANDORA manager in 2000, recalled, "We realised a significant amount of publishing was happening online, and nothing was being done to preserve it for the future, as we've long done with printed works."[73] Koerbin was pragmatic on the question of a 'dark age,' however, noting in our interview that "I don't think it was quite as dramatic as like . . . 'a digital dark age.' It was just all this material that, these publishing formats that we're not collecting."[74]

What made PANDORA significant for a national library was that the emphasis was not just on government documents and official publications but that it would also encompass "ephemera." Jenny Sinclair wryly noted that including ephemera meant that "a few of those annoying personal home pages that feature not much more than a photograph of the author's dogs and a list of their favourite celebrities" would end up in the National Library of Australia (NLA).[75] This was a big step. Koerbin remembers this being related to the 1996 Australian federal election, as the original planning had mostly involved serials and online periodicals. Yet with the election, held in March 1996 but campaigning before, the thinking changed:

> I think the penny started to drop really seriously with . . . the federal election in 1996, and a lot of the stuff was going online then, policy stuff. And I think that really moved us out of thinking, well, really, we're not just dealing with e-journals here. You know, there's a whole world there. And also, around the same time, a lot of government material started [appearing], government website set up. In fact, the National Library hosted a lot of that. So, we were very aware that this information wasn't [just] serials, wasn't [just] journals, . . . these were websites and material on websites that needed to be collected. So, I think that started to move our thinking more from just journals.[76]

After the 1996 elections came the later 1998 election and the subsequent lead-up and holding of the 2000 Sydney Olympics. Koerbin added that "it was [these] certain political and cultural events that played a part in really motivating development of our web archiving program."[77] Indeed, Hegarty echoes this in his work, illustrating the change between 1997, when the unit still thought of "publications" toward its shift of looking at "information" broadly defined.[78] Yet if it was to be broader than just e-journals, selection policies were needed. What to collect?

Between January and April 1996, the Selection Committee on Online Australian Publications (SCOAP) explored these questions.[79] Content needed to be Australian, and also needed to be solely online (i.e., not an online version of a print periodical) and of "cultural or historical significance to Australia."[80] SCOAP argued that the NLA should "select all publications judged to be of national significance" and pointed out that "publications that are not authoritative or do not have reasonable research value are usually not being selected."[81] As Koerbin notes, the key criterion was "is it of research value? . . . there were value judgements on it, to be honest."[82] The other key criteria were that sources needed to be *only* online. If there was a print edition, the web version would only be collected if there were significant differences.[83] This led to absurdities, such as spending hours on the phone with publishers to discover if something was online only or if there was also a print edition. As Koerbin (half) joked in our conversation, "by this time we could have archived the damn thing! You know, [it became] ridiculous that we're worrying about whether something is in print."[84]

Once SCOAP developed selection guidelines, the Electronic Unit was established in April 1996 to begin to identify and select titles, as well as cataloging guidelines. It was designed as a six-month pilot. Koerbin recalls that its goal was to "take these [SCOAP] guidelines and then go out and [collect] all the stuff that we wanted to do and that would be job done." Clearly these problems would require more than just a short-term pilot to solve, and accordingly in September 1996 the serials cataloging unit and the Electronic Unit were combined as an operational group.[85] Hegarty sees this six-month pilot as central to the evolution of the web archiving concept at the NLA, as it was when "the boundaries of the archived web lost [their] elasticity and began to stabilise as an object of knowledge, centred around the definition and management of the 'online Australian publication.' "[86] By November 1996, the name PANDORA was being used. And with it, the shift to emphasizing preservation and access, not just selecting and cataloging.

Enter one of the major challenges facing the NLA. Its legal deposit powers resided in the Copyright Act (1968). Without amendment, PANDORA could not use the authority of legal deposit. It would instead need to secure site-by-site opt-in.[87] Following consent, sites were then either captured on a one-time basis or continually. PANDORA staff then carried out manual quality assurance to ensure the preserved version was identical to the original. This was a time-consuming process that would dramatically improve the capture's fidelity.[88] While many site owners were happy to have their sites archived, others declined. "There's some early stuff that we really missed because we couldn't get the permissions," muses Koerbin, lamenting several "really important early sites" that were never preserved.[89]

Why did they not consider a dark archive, as per Sweden? Why not crawl first and sort it all out later? From its very inception, access was a key consideration. A December 1996 white paper explicitly connected preservation and access: "it is a matter of urgency that a strategy be put into place to ensure the survival of, and *continued access to*, these early electronic publications" (emphasis mine).[90] Koerbin explained why in part this was so important: "The objective has been open public access . . . we're not the Netherlands where you get on a train and be across the country in an hour or two. [Australia is a] huge country. So you couldn't say, well, you've got to come to Canberra. [laughs] . . . That was never an acceptable outcome for us."[91] In other words, "access has always been the principal driver."[92] While the library pushed for amendments to the Copyright Act, it did not wait for legislative change to provide access.[93] Years before the Internet Archive provided open access to its collections with its Wayback Machine, PANDORA hosted archived webpages via its library catalog. The exact date when archived web pages were made available via the catalog remains a bit opaque, but certainly by October 1997 there were indications that pages were available via the catalog.[94] "Anything we selected for PANDORA, we would catalogue," noted Koerbin, explaining the process by which records were linked to title entry pages in the catalog and made accessible to anybody (unlike the later Wayback Machine, which would connect archived webpages, each individual crawled website was a silo—"if there was a link to something else, even if that was in the archive, you couldn't go there [by a direct link in the archive].")[95]

The decision to be selective rather than comprehensive was a pragmatic one. Money was a factor, as funding came out of core operating budgets rather than having dedicated project funds. Yet the library was motivated by the goal of providing access to crawled content as well. "We were certainly aware of the Scandinavians . . . they were higher profile in doing domain harvesting," Koerbin recalls. "But I think it was just, we knew we weren't up for doing that sort of project here because of legal and technical constraints. But we certainly took an interest."[96]

Significantly, selectively crawling did not rule out changing strategies later. The library's approach was described by current director-general Marie-Louise Ayres as "radical incrementalism": practical steps with an eye on the future.[97]

While the idea and PANDORA formation date to 1996, websites began to be archived only in 1997. This delay had costs. In 1996, the team started working on selection guidelines and scouting. By the time crawls started, "some of the early candidates had already disappeared."[98] The digital dark age waits for nobody.

In October 1996, PANDORA's position paper "National Strategy for Provision of Access to Australian Electronic Publications" was released. It represented the NLA's official position and was widely distributed to national stakeholders in the library, cultural, and publishing communities. The paper recommended that legal deposit be extended to cover electronic publications, encouraged collaboration between libraries to select and preserve electronic information, called for an expansion of copyright to include the right to copy for preservation purposes, and argued that the NLA needed to "actively participate in the development of national and international standards for the discovery and retrieval of networked electronic resources."[99] It was a bold statement, setting the framework for PANDORA's operations.

PANDORA's criteria were selective yet broader in scope than the Canadians' had been. The subsequent December 1996 "Guidelines for the Selection of Online Australian Publications Intended for Publication" paper held that while the goal was to be as comprehensive as possible, the "potential volume of published material to be dealt with is overwhelming in relation to the resources available. A higher degree of selectivity is therefore necessary for online publications than is the case with print."[100] To be in scope, content needed to be "about Australia," be on a subject relevant to Australia if written by an Australian, or be written by an Australian "of recognised authority and constitute a contribution to international knowledge."[101] The paper also delved into ethically fraught issues, such as whether to collect "sites created for or by juveniles." On this front, PANDORA noted that such juvenile-related sites would be preserved only if they were popular, were official or semi-official, and had received acclaim or awards.[102] Despite musing about personal homepages, most such sites would not meet the selection criteria. PANDORA now had formal guidelines.[103]

The need for curatorial selectivity was baked into the project's foundations. Margaret Phillips, PANDORA manager, dwelled on this in 1999: "The National Library of Australia's pragmatic decision to take a selective approach was based on two factors. First, managing online publications is labor intensive and therefore expensive. Second, to the best of our professional judgment, many online publications have low current or future research value. We consider that it is better to identify those titles most likely to be of future research value and to apply our limited resources to managing them to a high standard."[104] Indeed, in 2003, a *Sydney Morning Herald* journalist deemed this selectivity to be PANDORA's distinguishing feature. PANDORA's "selective approach with clear criteria, cataloguing, search facilities, and staff who exercise quality control is one of two directions taken by national libraries," Lauren Martin wrote. The other, "mainly used by the Nordic countries, led by Sweden, is a whole-of-domain approach; it tries to harvest all net material from the country at a given time. This way is more comprehensive but less organised."[105] It was a good summary.

By 1998 other state libraries across Australia were also involved in web archiving. Some had created their own independent web archives, while others such as the State Library of Victoria had formally partnered with PANDORA.[106] This helped distribute the labor of selecting sites. As Koerbin explains: "The idea was that they would have their own selection guidelines and they would contribute to it based on their selection priorities. I mean, basically, they followed our selection guidelines and just wrote 'Victoria' rather than 'Australia.' So it was just a subset of [PANDORA and] wasn't that different. And their responsibility was really providing curatorial input."[107] The separate state web archives were also significant. For example, "Our Digital Island" was the web archive of the State Library of Tasmania. By 1999, "Our Digital Island" contained over 140 sites and represented the first full commitment made by a state library toward web preservation.[108] By the end of 2001, six other organizations were involved (the state libraries of Victoria, New South Wales, Queensland, South Australia, and Western Australia, plus the National Film and Sound Archive).[109] It continued to grow and by 2004, all of the mainland states (excluding Tasmania, which never joined) were contributing to PANDORA as well.[110]

Through these early web archiving programs, Australian cultural heritage was preserved in many unique and diverse ways. In 1999, the *Canberra Times* noted that webpages including humor sites, the children's band The Wiggles, and a website honoring the Australian spread Vegemite were all now being preserved in the National Library of Australia. The Wiggles should have been proud, as PANDORA was still quite selective in 1999. Margaret Phillips noted that there were then 214 preserved sites. While PANDORA had originally been focused on "scholarly, authoritative publications of long-term research value" they were "also interested in documenting how Australians used the Internet and what kinds of opinions, interests and concerns were documented on it."[111] Some sites understood inclusion as a badge of honor. As one humor site that focused on the "long history of failure and fiasco in Australian diplomacy" noted, they were "delighted by the National Library selection of its web site for lasting national significance and cultural value."[112] By 2000, 700 sites were then part of PANDORA, representing the beginning of rapid growth.[113] By April 2001, 1,100 sites, all accessible through the PANDORA website.[114] Three years later, in April 2004, there were over 5,000 sites.[115]

Given their active role in selection, curators faced controversy. In 2002, conservative Australian legislators got wind of PANDORA's plan to preserve pornographic websites. "Pornographic web sites are pornographic websites. They're not art and the National Library should be focusing on other things," argued a conservative parliamentarian. In response, PANDORA manager Phillips argued that PANDORA would "fail in its mission to record a broad snapshot of contemporary Australian life if it ignored the phenomenon."[116] A compromise was struck to collect it . . . but to then limit access to exclude minors. As one reporter put it, "if you prove your bona fides, you can even view a select number of turn-of-the-millennium sex sites though only while on a supervised computer in the Canberra library."[117]

Most of the Australian web did not meet the bar for collection, however, and would not be included. As Ian Morrison noted in 1999, this was a deliberate choice, as the "main reason [for this] was the insubstantial nature of most web publishing: the library does not attempt to collect printed ephemera comprehensively, and could see no justification for collecting its electronic equivalent."[118] In other words: "This is the key point about selectivity: it is our normal practice."[119] Even if it was possible, the Australians argued, collecting too much information would lead to an undesirable outcome. When it came to print material, they did not collect everything. Why should electronic documents be any different?

Morrison perhaps overstated the case against collecting ephemera. While the National Library of Australia had always been selective, it did have an ephemera collection dating back to the 1960s "as a record of Australian life and social customs, popular culture, national events, and issues of national concern."[120] Koerbin noted this in our interview as well, recalling this edict against ephemera as "a little odd, too, because the National Library has a long tradition of collecting print ephemera and a very fine collection, and so understands the value of that."[121] An added issue, of course, was the pragmatic challenges of collecting ephemera when you needed to find the copyright holder to give you permission to collect it.

An increasing downside of PANDORA being early to web archiving was that, by 2003, its original 1996 selection criteria were lagging the reality of what the web had become. New types of content were now commonplace. One journalist lamented that PANDORA was forced to "concentrate on the limited areas of significance chosen in 1996: government publications, academic journals, conference papers, e-journals, and topical sites," while webcam feeds, blogs, chat transcripts, games, and databases were out of scope and thus uncollected.[122] In 1999, Morrison argued that while selectivity may have been normal practice, there were increasing reasons to expand its mandate to encompass ephemera. As he noted, libraries such as the NLA and its state counterparts "maintain large ephemeral collections within specific areas; but neither attempt to collect every political leaflet or sale catalogue produced. Both aim to preserve a solidly representative sample of the types of material disseminated."[123] Koerbin recalls: "Around 2000, 2001, 2002, we were certainly looking at domain harvesting. We did have a project to look at . . . the feasibility of it, because we are certainly aware of what the Swedes and the Scandinavians generally were doing, you know, taking that broader approach. So, it wasn't like we had an ideological fixation [laughs] on doing selective [crawling]. It was really because of our incremental, ad hoc approach to it, to the whole thing that that was the way to do it."[124] This feasibility study, the Australian Web Domain Harvesting Project, was carried out in mid-2001. It never proceeded to full implementation at least "probably due to lack of maturity of tools available."[125] By 2005, the National Library was asking the government for more comprehensive legal deposit legislation to support a broader web archiving approach.[126] Without copyright and legal deposit reform, there couldn't be access. Access was so ingrained as a core principle that a dark archive was not under consideration.

An early move toward providing access to material collected in such mass harvests came in 2006, when an amendment to the Copyright Act was passed. This new section provided for the use of copyrighted work by libraries and archives for socially beneficial purposes.[127] Such use needed to comply with Article 13 of the Trade-Related Aspects of Intellectual Property (TRIPS) Agreement, which—as Koerbin recalled—made "it tricky to apply and somewhat opaque in respect to the web archive, although we always understood the intention of the clause was to allow us to do such things as making the web archive openly accessible."[128] Ultimately, this opaqueness meant that the ostensible powers granted in 2006 were not used until a subsequent amendment passed in 2016.

In any case, selectivity was showing its age. Perhaps the Swedish and Internet Archive approach would be more appropriate? Could the success of the NLA be leveraged to make it a broader collection? Wendy Smith noted in 2005 that PANDORA was continuing "to emphasize 'print-like' objects . . . at the risk of failing to capture the full range of uses that the web now has in communication and dissemination of information of all sorts, and failing to accommodate the evolutionary, increasingly database-driven, steps of the web."[129] Smith pointed to the work of Axel Bruns. Bruns, an Australian New Media scholar, had presented a paper, "Contemporary Culture and the Web," to the National Library of Australia. In it, he outlined the web's evolution. Bruns sketched out a "Tale of Two Webs," noting that the web was moving "towards more user co-creativity." This could be seen in a shift from homepages to blogs, community news to open news, and resource sites to wikis.[130] A transformed web needed a new collections strategy.

Yet much of this would need new legislation.[131] As noted above, the 2006 reforms did not change operations much at the NLA, in part as there was not yet the appetite for the risk the institution might face.[132] As Smith explained, "in the decade since those [selective acquisition] criteria the nature of the web has evolved and the patterns of its use have changed, signaling a need to extend the selection criteria."[133]

If in 2006 the Copyright Act had been amended in theory to facilitate greater access, in 2016 it was amended to extend legal deposit to digital formats. Yet even before that there was a new "willingness to manage risk since the legislation at the time was ambiguous."[134] PANDORA began crawling the Australian web domain in the summer of 2005.[135] As journalist Steve Meacham wrote in the *Sydney Morning Herald*, this was a significant shift from PANDORA's selective origins: "who's to say what should be collected? Which of us can predict what future generations will find culturally significant?"[136] The Australians were adopting the model that the Swedes and the Internet Archive had adopted in 1996.[137] One downside is that this material was inaccessible until March 2019, when PANDORA was subsumed as part of a much broader Australian Web Archive (which brought together government sites, PANDORA, and the broad domain harvests).[138] A time-limited dark archive had carried the day after all.

Yet by exploring the com.au top-level domain, the PANDORA project team could understand the scope and scale of the Australian web domain. Beginning in June 2005, the National Library of Australia ran a four-week test crawl with the Internet Archive. As noted in chapter 3, the Internet Archive had by then developed its own crawlers, independent of Alexa Internet, thanks to the development of the Heritrix crawler. Using Heritrix would thus generate crawls that were much larger than any done previously.[139] The test crawl lasted six weeks, encompassing 185 million websites and compiling 6.7 TB of data.[140] Koerbin also noted that this was mainly possible due to Heritrix: "Why 2005? Well, it's pretty simple, actually . . . Firstly, that's when the Internet Archive released really a functioning version of Heritrix [a web crawling software]. And so they were in the market to do this large-scale harvesting. So here was someone we could go to. But also at that time . . . [t]here was money [available for this purpose] . . . So, this became a possibility that we could then contract the Internet Archive. [It] had to be framed as: 'This is a collection. We are buying this collection item called the Australian Web.' "[141] Everything soon came together, launching a model that national libraries would soon use across the world, including the Library of Congress and Library and Archives Canada.

One difference arose with this newly collected material. It would be complicated to provide access to it, as the NLA had not used the opt-in model. The Copyright Act would not be amended until February 2016. Yet broad collecting was being done on a scale that precluded permissions. Accordingly, as with Sweden, Australia now had a dark archive. This was a dramatic shift for a web archive defined in terms of both selectivity and openness. Yet, such a shift was perhaps inevitable. Many other libraries would follow suit.

For the web had changed, meaning that web archiving had to as well. By 2010, the web was more complicated and its importance to the broader cultural record was clear. Koerbin noted that while Australian web archiving had begun with the assumption that the web was a publication, it had dramatically evolved: "But once it started moving into being as much a communication medium as a publishing one, the line between private and public [was] not so clear. In the early days, people may have had a sense they were publishing. Now, people just contribute or get on to a social networking service—they are not thinking they are publishing. A lot of people would have the sense they intend what they put up online to be private or ephemeral. They are naive in thinking that."[142] PANDORA itself changed to meet these challenges. Beginning in 2011, the web archive began to collect Australian government websites under the auspices of the Australian Government Web Archive. The Copyright Act revisions of 2016 enabled collection, following on the 2006 amendment, which had enabled access. Crucially, the 2016 changes helped catalyze access as well—as Koerbin recalls, "the appetite for risk in using [the provisions of the 2006 amendment] only emerged following the introduction of legal deposit for online materials in 2016."[143]

This all came together in 2019, when PANDORA was subsumed into the broader "Australian Web Archive," which brought together PANDORA, the Australian Government Web Archive, and the large domain crawls.[144] Crucially, all nine billion files within that combined web archive were available to the public through a custom front-end interface with full-text search. The web archive was also integrated into the National Library of Australia's broader Trove web interface.

PANDORA inspired other national libraries. The Library of Congress looked to the Australians. Even as Australia moved toward a comprehensive approach, its initial success with a selective framework would lead to long-reaching impacts felt well into the twenty-first century. The Australian experience is also a useful entry into the shifts that national library web archiving faced. Initially understood as something akin to backing up electronic journals or a few significant sites, as society shifted online, the value of what had previously been seen as akin to ephemera became increasingly important. Such material eluded selection because of its scope and scale, necessitating new approaches to partnership, copyright, and even legislation. As countries such as the United Kingdom and France evolved their own programs, comprehensiveness would come to dominate. The National Library of Australia's experience bears this out.

MINERVA: The Library of Congress

The Library of Congress's web archive traces its origins to October 1998, when it received a donation of an Alexa Internet web crawl. In a nod toward late-1990s physical understandings of the web (and some theatrical flair), the donation came in the form of a sculpture. The sculpture, "World Wide Web 1997: 2 Terabytes in 63 Inches," had been crafted by American artist Alan Rath. Four stacked computer monitors, attached to forty-four digital archival tapes, intermittently flashed selections from the almost half a million archived pages.[145] This was not its first public appearance, as the sculpture had previously graced the stage at the Getty Institute's Time & Bits conference. It was an evocative representation of the archived web. The Library of Congress was thrilled with Rath's statue and the accompanying data. Winston Tabb, associate librarian for Library Services, explained that Alexa's donation would ensure "that one of the most significant collections of human thought and expression born of a new medium is preserved in the national collections."[146] Congresswoman Zoe Lofgren was equally impressed. She noted that Alexa's donation would mean that "the Web's valuable but fleeting resources will be retained for generations to come."[147] A new chapter of the Library of Congress's long history had begun.

The sculpture inadvertently illustrated the challenges facing the Library of Congress's web archiving program. Rath's sculpture spent time in a public place, before breaking and being moved to the web archiving team's offices. Finally, it would later be returned to the Internet Archive to return to public display (where it remains today). The crawl itself, however, was never actually added into the Library of Congress. As Abigail Grotke, who was then with the web team and would later lead it, "the content itself we never got . . . it was on these hard drives." While the team briefly considered extracting the data before returning the sculpture to the Internet Archive, that never came to pass.[148] The technical and permissions issues were insurmountable.

One of the emerging requirements of being a de facto national library was and is the stewardship of digital cultural heritage.[149] Given the United States' outsized role on the internet and the web, this was especially pressing. In light of emerging web archives in Canada, Australia, and Sweden, it was clear that the Library of Congress needed a stronger digital acquisitions strategy. This was vividly illustrated by a 260-page report released by the National Academy of Sciences in July 2000.[150] *LC21: A Digital Strategy for the Library of Congress*, commissioned by the then-Librarian of Congress James Billington and produced between 1999 and 2000, provided an essential overview of the challenges facing the Library of Congress in the digital age. Hedstrom recalled *LC21* as a "really fundamental" report alongside the earlier (and aforementioned) Task Force on Archiving of Digital Information.[151] *LC21* provided a sobering overview of where the institution was at the dawn of the twenty-first century. While the Library of Congress was still arguably "the largest and most prestigious library collection in the world," *LC21* highlighted that it was significantly lagging in "receiving and archiving the born-digital product of the nation." A "truly functional contemporary Library" would require adopting new processes for taking both physical and digital objects, from books to CD-ROMs to webpages. Neglecting this would run the very real risk that "the Library of Congress could become a book museum."[152]

The report focused at length on web archives, appropriately looking to Canada, Sweden, and Australia for inspiration.[153] Web publishing challenged the traditional publishing world that the Library of Congress operated in. *LC21* noted that some of the webpages and other digital publications promised to be "as important as records of current research and creativity as were the journals and books of the eighteenth, nineteenth, and twentieth centuries."[154]

The recommendations make for sobering reading today as they address issues that still remain unsettled. First, *LC21* argued that the Library of Congress needed to clarify its legal deposit powers for websites. If it needed more power to meet its goals, the Library of Congress needed to seek legislative authority. The report also recommended that the library conduct web archiving pilot projects, with an eye to later translating these pilot projects into a broader strategy to preserve the web. Selectivity emerged as a critical theme for *LC21:* "The committee believes that LC needs to be not only more ambitious but also more selective in the methods it uses to build digital collections. Increased attention to selection is necessary because of the explosion in digital information of widely varying quality and interest. The Library cannot be expected to collect everything, especially if collecting carries with it an obligation to preserve what it collects . . . By comparison, the world of Internet publishing is new, raw, and wild. New models for defining Library collections require a redefinition of what constitutes a library's 'collection.' "[155] Part of the issue facing the Library of Congress was the web's dramatic growth. If one could imagine the finite boundaries of the web in 1996, by 2000, the size of the web was beyond definition. "If you accept that the Internet should be collected, what does that mean?" asked an Oregon library director who had been on the *LC21* committee, "Today's Internet? Tomorrow's? all? And do you collect just the first layer of a Web page and none of the links? Three layers? Ten?"[156] The stakes were clear. Ephemeral data meant there would be no second chances.[157] Data not collected today might not be there tomorrow.

Others were less bullish on the *LC21*'s potential to make an impact. Hedstrom recalled a group of *LC21* authors gathering to chat about it, and how they began to riff on a meme: "it was when Bill Clinton was running for re-election in '96, where he was saying 'it's the economy, stupid' and we were going, 'it's born digital, stupid,' because the focus of the LC had been on digitization . . . But the big problem is born digital."[158] Part of this was reinforced in a meeting with the Librarian of Congress, where as Hedstrom recalled, "it was very clear to us that [Billington] didn't understand at all the difference between something that was digitized and born digital."[159]

But *LC21* gave impetus for pilots. As Hedstrom notes, it at least acknowledged "the born-digital business."[160] Throughout summer and fall 2000, the Library of Congress carried out its web archiving pilot project: MINERVA, or the "Mapping the Internet Electronic Resources Virtual Archive." MINERVA was a backronym, similar to PANDORA. Named after the mosaic of the Roman goddess Minerva in the great hall of the Library of Congress's main Jefferson building, MINERVA was the Library of Congress's first entry to the world of web preservation. The project focused on the ongoing American presidential election campaign between George W. Bush and Al Gore. As that election extended past voting day into the Florida recount and arguments at the Supreme Court, it was a consequential example of why web archiving mattered. These sorts of focused collections would define the MINERVA project until around 2005, when it would move beyond elections, events, and Congress (House and Senate content began to be collected in 2003).[161]

The first pilot consisted of two main initiatives: the in-house collection of a small number of websites to understand "practical issues," as well a partnership with the Internet Archive to capture "some 150 to 200 web sites on a daily basis."[162] This required decisions on whether the project would approach its pilot from comprehensiveness or selectivity. Refusing to see this as a stark binary, the interim project report noted that "there are strong reasons, both scholarly and economic, for considering bulk and selective collecting as two facets of a single strategy."[163] MINERVA refused to buy into the false dilemma of selectivity versus comprehensiveness.

While the in-house pilot focused only on WhiteHouse.gov as well as the Al Gore and George W. Bush campaign sites, the experience of the team led to reports that laid a foundation for the coming two decades.[164] From its inception, MINERVA considered the problem of access (replay versus raw source files; on-site versus remote), sustainability, comprehensiveness or selectivity, and beyond. The September 2001 final report made a case for a flexible model based on collaboration. Both the interim and final reports, written by William Arms, were impactful. As Grotke recalls, "a lot of what we were [later] trying to establish was based on [them]."[165]

This was an early example of a model that would underpin the Internet Archive's global role. These partnerships between national libraries and the Internet Archive would see curation done by the former and technical aspects by the latter. The report explained: "As a Silicon Valley insider, the Internet Archive has access to advanced technology and expertise. Web crawling, automatic indexing of web sites, and the techniques of managing massive collections on commodity hardware are well established among the best Internet companies, but have not percolated into the mainstream computer industry. Consequently, the Internet Archive's costs for collecting and preserving web materials are much lower than the Library's would be."[166] Such a partnership was mutually beneficial. "The converse of flexibility is instability," *LC21* argued. The Internet Archive was indeed more flexible when it came to its operational latitude. It could quickly hire, adopt new technologies, and in general bring a versatility that a government organization could not. But in a century the Library of Congress or other national libraries had a higher likelihood of still existing than the Internet Archive.

There was, perhaps, a more cynical equation at play too in the relationship between the Library of Congress and the Internet Archive. It was, as Kahle noted, at least somewhat "a relationship where we could do things they couldn't do, or they didn't feel that they could do." He expanded on this: "they had plausible deniability, so if we got shot up then, you know, they wouldn't get dragged into the lawsuits of whatever, but we [could] also kind of leverage their patina." Kahle and the Internet Archive could "do things that were a step ahead of where they thought they could be.[167]

It was a valuable partnership for both parties. Being based in the collections unit, rather than IT, MINERVA relied on the Internet Archive's technical expertise. It was not without hiccups: the Library of Congress staff struggled with the proprietary nature of the Alexa Internet crawls (Grotke recalls a colleague "digging around and pestering them with questions because she was the first person to ever do a quality review") as well as problems around missing images and stylesheets given the importance of high-quality fidelity to the library.[168] It was only later, as with Australia, when the Internet Archive began doing its own non-Alexa crawls with Heritrix that the Library of Congress could better understand the archiving process.[169]

Crucially, however, the framing adopted by the MINERVA team broke down the false binary of comprehensiveness versus selectiveness. The final project convincingly demonstrated how the two approaches were complementary. Bulk collection needed to be scoped by a skilled curator or librarian, making the distinction fuzzier. "For example, a librarian might define a category of material by some set of criteria," the authors noted, adding that the librarian would then "rely on automatic processes to identify the web sites that satisfy those criteria and collect them."[170]

The Election 2000 web archive collected by the Internet Archive launched in June 2001: 87 million pages, over a thousand distinct websites, two terabytes in total, released via a special Wayback Machine portal.[171] "Election 2000, as It Happened," declared the Library of Congress's webpage. The collection's significance was stressed by Winston Tabb: "This was the first presidential election in which the Web played an important role, and there would have been a gap in the historical record of this period without a collection such as this."[172] The media was wry. "Though no one may care to relive Election 2000," a columnist wrote in *USA Today*, "this digital archive . . . offers a 'Wayback Machine' to let you see how that tumultuous election season played out online."[173]

We can use the Internet Archive's Wayback Machine to see what the (now defunct) Election 2000 Wayback portal looked like. Sites were organized by political party, news, House, and Senate sites, as well as sites on a variety of topics such as state-level vote trading, encompassing the period between 1 November 2001 and January 2002's Inauguration Day. As a disputed election, this was a consequential record of a fraught period in American political history. Today, the collection is split across several different web archive collections hosted by the

Library of Congress: the general collection, the public policy web archive, and the elections website. This reorganization highlights how digital collections continually change.

The Library of Congress today remains the epitome of a selective web archive, which is notable for a country as large and significant as the United States. Unlike England or France, for example, the Library of Congress does not do comprehensive domain crawling. This is in part because of the library's unique status (as well as the absence of a defined national domain). While it has many of the de facto powers of the national library, legal deposit—enshrined as "mandatory deposit" under the Copyright Law of the United States—has not been expanded to include websites. Until 2002, MINERVA used a notify process rather than a strict opt-in permissions process. This would have considerable impact during the 11 September 2001 event web archiving efforts.

Yet, by 2002, Library of Congress management ensured that the web archiving program both took these questions seriously and developed a permissions process.[174] Indeed, Schneider and Foot—who had collaborated with the Library of Congress around 9/11 collecting and subsequently the 2002 election archive, recalled several issues that sprang up around copyright on the very morning of launching the web archive at a public event (Schneider recalled "there were roadblocks to that event that morning, if I remember correctly").[175] Thinking back on that event, Grotke recalls that "because of the event there was more attention paid to the web archiving program, and a number of discussions around the legal/permissions approach were happening internally, which led to coming up with a plan to ask permission."[176]

The Library of Congress continues to collect a large amount of information, and indeed, as a petabyte-scale web archive that collects hundreds of terabytes a year, it is crawling more data than any other national library. Unlike other national libraries, and perhaps because of its historic and expansive mandate—which sees extensive collecting beyond the borders of the United States—it also has a notable international perspective to its collecting.[177] Collecting includes government webpages and electoral processes from across the world, including large collections from Africa, Asia, Latin America, and Eastern Europe. Scholars looking to understand elections in countries as varied as Tajikistan, Afghanistan, Brazil, the Philippines, or Bangladesh may look to these curated collections.[178]

This expansive collection scope is facilitated by the library's approach. Crucially, the Library of Congress ignores robots.txt and collects material from websites even if a webmaster does not want it to be collected, but as a compromise, such material can then only be viewed in the reading room. Fortunately, here is one area where the occasionally American-centric nature of the Internet Archive's collection strategy can help: the Internet Archive's comprehensive approach complements the selective work that happens at the Library of Congress.

As many of the world's national libraries move toward emphasizing comprehensiveness, there is value in building strong selection capacity. In 2020, on the twentieth anniversary of the LC's web archiving program, the *New York Times* argued that the LC demonstrated the value that a small, dedicated team brought to capturing ephemeral sources such as internet culture and memes. Through collaboration with institutions like the American Folklife Center, websites that composed the "Web Cultures collection"—web comics, independent music artists, and the like—are now preserved in perpetuity. The article highlighted some of the issues with not having a comprehensive legal deposit policy: "Megan Halsband, who oversees the Webcomics collection, still mourns the death of Joey Manley in 2013, and with him, the influential sites he published like Serializer and Girlamatic. And she has so far been unable to archive another popular webcomics site, The Oatmeal, because in that case, the cartoonist who runs it has never responded to her emails seeking permission. (The library has an opt-in policy.)"[179] Complemented by the Internet Archive—which has (fortunately!) archived the Oatmeal—the digital heritage of the United States will be a well-documented one.

Much of the recent shift in the Library of Congress was driven by the terrorist attacks of 11 September 2001, which we will return to shortly. The attacks of September 11th, 2001, led to such wide-ranging impacts on American culture, media, society, and politics, that it spurred a reassessment of how digital information should be documented.

Conclusion

Legal deposit, the obligation for publishers to deposit published materials to national libraries, sat uneasily with the broader world of the web, where everybody could be a publisher. The country-specific mandates of national libraries complicated all of this further. A global network sits unevenly with national-level mandates and institutions. As the Canadians noted during their early EPPP pilot, hyperlinks span national borders and do not lend themselves to nice and tidy "publications" that can be neatly cataloged and preserved.

Expanding the concept of legal deposit to the web raises several questions. First, scale. Just how many websites are published in a country, for example? And how to find them, especially those published outside of a given-country-level, top-level domain (the ones on .com or .nu, as the Swedish discovered, not just .se).

Second, questions of capacity, or an understanding of the required storage and technical infrastructure. Third, copyright and ethics are key. Unlike a publisher requesting an ISBN, a website's author might be a 12-year-old publishing a website about a board game. Comprehensive crawls by national libraries, however, have the advantage of being stewarded by long-standing government institutions. Even if concerns around the long-term sustainability of the Internet Archive are occasionally overwrought, the long-term prospects of the British Library or the Library of Congress existing are excellent.[180] With legal deposit often came the switch toward comprehensiveness, something that many national libraries saw as impossible without the legal foundation.

The United Kingdom's British Library presents a good example of this shift. While the British Library is today a significant global web archiving leader, it emerged on the scene later than the Canadians, Swedes, Australians, and Americans. The British Library's original web archive, started in 2005, was a selective one. While in 2003, the United Kingdom expanded legal deposit to include websites, this legislation still required authors to opt in to collection. As one can imagine, while many webmasters were fine with their sites being archived, many more were dubious or they saw invitations as spam. Between 2005 and 2013, the British Library created a curated collection of websites, which—thanks to the permissions process—can also be made openly accessible for research. In 2013, legal deposit legislation was changed to allow the widespread crawling of the British web without needing individual consent. The British Library then began to harvest all websites published in the United Kingdom, from blogs, homepages, activist social media streams, large corporate websites. It could also ignore robots.txt and thus steward this material for generations.[181] However, the collection now had additional regulations, notably around access (onsite only) and limits on simultaneous access to crawled resources. Today, the British Library's web archiving team combines curatorial expertise with technical ability and leadership. Large annual comprehensive crawls are complemented by selective harvesting, further illustrating the importance of moving beyond the false dilemma of selectivity or comprehension.

Other national libraries would follow. The Bibliothèque nationale de France gained web legal deposit powers in 2006, subsequently strengthened and clarified five years later. Iceland did as well in 2003, followed by Denmark in 2004. Over the past two decades countries as varied as Austria, Croatia, Estonia, Finland, Germany, Portugal, New Zealand, Norway, Spain, and Slovenia have all either enacted new legislation or worked within existing frameworks to ensure the comprehensive collection of web content in their domains.[182]

Between 1996 and the early 2000s, web archiving grew from a niche concern for a small handful of national libraries to a core concern for a global network of libraries. This was best seen in the July 2003 founding of the International Internet Preservation Consortium. The IIPC brought twelve libraries (Australia, Canada, Denmark, Finland, France, Iceland, Italy, Norway, Sweden, United Kingdom, United States, and the Internet Archive) together to "work on developing standards, tools and policies to help acquire, preserve, and make accessible knowledge and information from the Internet for future generations everywhere, promoting global exchange and international relations."[183] Its three goals were to "enable the collection of a rich body of Internet content" to be preserved; to foster development and use of common tools and standards, and crucially, to encourage "national libraries everywhere" to take up the cause.[184] While individual activities wax and wane, the existence of such an international forum illustrates that web archiving is a basic and fundamental function of a modern national library.

In 1994, when the Canadians started a small pilot project to harvest web-based electronic journals, few foresaw how dramatically the library landscape would evolve over the coming decade. The Internet Archive was but a glimmer in Kahle's mind when those first Canadian websites were preserved. A process had started, which would see debate over how national libraries could sustainably steward their nation's digital cultural heritage. While the Swedes would adopt a different approach, the debates, discussions, and archives that took shape over the coming years would both preserve culture and help memory institutions learn how to do so.

All of this activity, too, laid the foundation for the ability to document rapidly evolving events. Web archives were used to deal with scheduled events, such as elections or sporting events. But how quickly could they adapt to a changing world? During the terrorist attacks of 11 September 2001, and the tumultuous global aftermath of the day's carnage, many institutions would face the epitome of a rapidly evolving event.

CHAPTER FIVE

Archiving Disaster

The Case of 11 September 2001

The terrorist attacks of 11 September 2001 against the United States were among the first major disasters of the web age. History seemed to be made in the minutes, weeks, and months after the first plane hit the World Trade Center's North Tower that morning. As phone networks failed, many turned to the web and email to ask about the safety and security of friends, family, and colleagues. Office workers, working in cubicles and offices away from television, tried to make sense of the day on quickly failing online news sites.

It was quickly clear that the events of that day would dramatically change both global politics and the lives of many around the world. Within hours of the attacks, archivists and librarians were capturing webpages, television feeds, voice mail messages, and email messages. Within weeks, new and emerging forms of memory websites were created, allowing anybody to record reactions, thoughts, and details about how their lives had been transformed. A web portal was launched a month later, providing immediate and indexed access to the archived web.

This all happened quickly. Just six years earlier, the specter of a digital dark age had haunted those worried about the future of our collective cultural heritage. By 2001, the situation had changed. The Internet Archive was then over five years old. The Library of Congress's web archive had just celebrated its second anniversary. All of that infrastructure, put in place to record either long-foreseen elections or the everyday evolution of web culture, suddenly needed to preserve the very specific and ephemeral online experiences of how people understood and responded to the attacks. The Digital Dark Age of 11 September 2001 would be measured not in months but rather the hours that elapsed between event and the beginning of active preservation.

The attacks vividly demonstrate how the digital dark age was, in broad strokes, averted. If before 1996, observers were anxious and worried about the short life span of webpages, the challenges of 11 September 2001 vividly demonstrated how memory institutions could effectively respond. Without the infrastructure, expertise, and web archiving programs that had been set up over the previous five years, the record of the attacks would have been radically circumscribed. It was clear in the day's aftermath that the cultural record was now increasingly mediated through the web. Collecting this material would be just as critical as accumulating the print record. Few questioned this, underscoring the generally accepted importance of digital preservation.

Through this monumental collecting effort, web archiving and grassroots digital collection came of age. The scale and speed of the collecting effort would lead to dramatic challenges, met primarily through considerable investment. The heritage sector rose to the challenge before it. There would be no digital dark age of the attacks.

The 11 September 2001 Attacks and the World Wide Web

At 8:46 A.M. Eastern Time, American Airlines Flight 11 was flown into the World Trade Center's North Tower in New York City. That was the first disaster of many that day, part of terrorist network Al Qaeda's coordinated series of terrorist attacks against the United States. Three minutes later, CNN cut from commercials to a live shot of the burning tower. Viewers who were watching live soon saw United Airlines Flight 175 crash into the South Tower at 9:03 A.M. "This looks like it is some sort of concerted effort to attack the World Trade Center that is underway in downtown New York," remarked the alarmed anchor on ABC. At 9:41 A.M., reports of another flight, later confirmed to be American Airlines Flight 77, crashing into the Pentagon in Washington, DC, began to break across television networks. The South Tower collapsed at 9:59 A.M., the North Tower at 10:28 A.M., and it was clear that the casualty count would be in the thousands (coverage over the first few days worried that it might be an order of magnitude higher). United Airlines Flight 93, where passengers fought back against the hijackers and a struggle ensued, crashed at 10:03 A.M. Live coverage from that crash site in Shanksville, Pennsylvania, would not appear until 12:42 P.M.[1] Much of the world ground to a halt as the significance of what was happening became clear.

Word quickly spread. In the United States, 44% of Americans learned about the attacks from a television (either live or as they woke up), 22% from radio, and 31% from other people (as researchers noted, this was "possibly because the attacks took place during a time of day when many people [on the east coast] were just congregating at their workplaces and probably because of the magnitude of the news."[2]

Many Americans turned to their televisions as the main way to follow the unfolding events of the day. Office workers who remained at desks without ready access to television used the web to follow the news. The allure of television was clear. Evocative visuals were being broadcast from New York and Washington, in a way that eclipsed the limited streaming capacity of news websites (even before web loads began to crash the sites). As Pew Research noted, "even the most wired Americans were wedded to the television and the telephone after the attacks." Some 79% of Americans (80% of heavy internet users) considered the television to be their main source of information. Conversely, only 2% of all Americans, and 6% of heavy internet users, considered the internet to be their main source of the day.[3] Television filled immediate news needs. Across the world, wall-to-wall coverage on television provided running commentary. Coverage was often panicked and confusing, given the sheer amount of information and events that seemed to be happening all at once. During these confused hours, the web did not play a central role in how people initially understood the attacks.

This was compounded by web service failures. Despite its Cold War origins and distributed nature, the internet did not seem terribly robust to the average web user that morning. That said, the main backbone that supported the internet remained operational and proved to be resilient. This was significant given the disruption of crucial physical internet infrastructure in lower Manhattan: the collapse of World Trade Center 7 later in the day caused severe damage to the Verizon Building, including its underground telecommunications vaults and critical internet connections. As New York City was a global internet "supernode," this led to a limited amount of global degradation. For example, a considerable amount of intra-Europe and intra-Africa traffic routed through New York City. This meant that networks in Italy, Germany, Romania, and South Africa were all affected. Ironically, some overseas users were more affected than those a few miles uptown in New York City.[4]

Servers hosting news media sites began to fail under the surge of users.[5] Major websites, such as CNN.com, NewYorkTimes.com, and Yahoo! news, were effectively unusable for several hours after the attacks. This was an issue of too much demand. Upon learning of the second plane impact, CNN's web director "stood up in his cubicle and shouted to other staff members to take steps (such as bringing extra servers online) to prepare for an increased demand for news. By the time he sat down, that spike had already arrived."[6] Nothing could be done on such short notice to make even a well-resourced site like CNN available to such an onslaught. Writing the next day in the *Detroit Free Press*, a technology columnist lamented that the "Internet failed miserably in the hours immediately following yesterday's terrorist attacks."[7] As main news sites crashed, many flocked to other sites—including unofficial New York City Fire Department websites or police scanner feeds—until those also collapsed. Users were like a flood, seeking information and bringing down websites as they arrived en masse. Record user numbers were reached throughout the day, which does not even include the many more people who could not reach the sites at all and stopped trying. CNN.com saw 162 million pageviews throughout the day of the attacks, for example, compared to a normal 14 million.[8] As part of the response, news organizations radically altered websites: stripping away advertisements, images, and superfluous content. CNN shrank its homepage from 255 KB to 20 KB, whereas CBS hosted only one story and one photograph.[9]

Other web sites were able to meet the demand for information, however. Technology website Slashdot, a place where users would submit a link to a news story with commentary to foster online discussion, provided short, succinct updates on the unfolding events. This happened elsewhere too as many high-capacity technology sites made sure that their readers, both regular and novel, could understand what was happening. That said, as Lawrence Lessig recalled of the events of that day and its immediate aftermath, the provision of news on the web had a distinct flavor: users "constructed photo pages that captured images from around the world and presented them as slide shows with text. Some offered open letters. There was anger and frustration. There were attempts to provide context. There was, in short, an extraordinarily worldwide barn raising." As he put it, there "was ABC and CBS, but there was also the Internet."[10] The web would play a significant role during that day, and in doing so would showcase its ability to connect users not just with information but with *each other*.

Where the web thrived was not in the transmission of information from news sites but in the facilitation of communications *between* people. Worried users wanted to check in on loved ones and make sense of the unfolding attacks. Telephone networks were overloaded in New York City, Washington, DC, and to some degree in Boston (when it became clear that two hijacked planes had departed from there). The volume of calls led to difficulties in calling in or out of the cities, particularly New York. At the attack's peak, 92% of cellphone calls failed in New York and 56% in Washington.[11] Approximately a third of Americans encountered difficulty using a phone at some point on that day. This was an unprecedented network collapse.[12]

With phones failing, many turned to the internet to connect with friends and family. As Tom Scheinfeldt, who I will introduce shortly as director of a major digital archive of the attacks, noted that "in the past, when people reacted to these kinds of events, they went to the local bar or got on the telephone. But the telephone system went down and people were reacting from their desks."[13] Here the internet's resiliency and redundancy carried the day. As the National Research Council of the National Academies noted in its retrospective on the internet's technical performance that day:

> First, the Internet degrades under load more gracefully than does the voice network. If sufficient capacity is not available, the cell-phone network will not permit new calls to be set up. In contrast, the Internet makes use of mechanisms that continue to accept new messages but reduce transmission rates when the network is congested. Also, by virtue of their flexible design, Internet-style communications lend themselves to human actions that reduce the load—whether by substituting a brief text message for a data-intensive voice call or removing data-intensive graphics from a Web page (as CNN did in the face of high loads).[14]

If the internet did not fully meet the needs of information seekers, it certainly facilitated communication.

For many people, this included sending an email to a loved one making sure they were okay or mailing their contacts to let them know they had not been a victim of the attacks. In the days that followed, over 100 million Americans sent or received "messages of emotional support, messages of concern for others, or questions about victims of the attacks," according to a UCLA study.[15] As internet scholar Jeffrey Cole noted, "Sept. 11 was the first major national crisis since the beginning of the Internet and e-mail . . . in the aftermath of the attacks, e-mail had a profound influence on how Americans communicate."[16] Yahoo! and America Online saw "above average" use of their instant messaging systems, and dedicated listservs quickly appeared for people to discuss the attacks even as they were unfolding, including Sept11info and WTCattack.[17] Writing the day after the attacks in the *Washington Post*, reporter Joel Garreau wrote of the "shaken global village," noting that even if the news sites failed, "when it came to many-to-many connections, the Internet showed what it was made of . . . it morphed and evolved to meet its users' needs. Suddenly, it was a lifeline for loved ones and a public forum for those who needed a place to yell their cries for vengeance or simply wanted answers."[18]

Specialized services quickly emerged to overcome communication shortfalls. Prodigy launched an "I'm Okay Message Center" at okay.prodigy.net on the afternoon of the attacks. Users could add their names to a national "I'm Okay" list or search for people via region (New York, Washington, Pennsylvania) and the first letter of their last name. There were several of these overlapping services, testament to widespread anxiety and loss. Within three days, a site run at Berkeley pulled data from the other leading safety check-in services and aggregated over 30,000 check ins.[19] Across the internet, websites responded to the attacks by providing links for blood drives and charities, as well as showing patriotic support by embedding American flags on their websites. Many of these efforts would last for months.[20] As websites changed, so did user behavior.

In the following weeks and months, the internet and web were used in new ways. The inclination of some users to repeatedly refresh news sites and to send instant messages to friends and family to make sure they were okay led to lasting behavioral changes. In the week after the attacks, 42% of internet users went online to read the news. Network loads soon stabilized and websites returned online.[21] Surveying Americans a year later, Pew Research noted that some 19 million users had connected with "family members, friends, former colleagues and others that they had not contacted in years," with 83% of them continuing to keep in touch.[22] Overall, citing the terrorist attacks as the major reason for their changed behavior, over the year that followed two million more Americans (15%) used email, 10 million more Americans (43%) got news online, 4 million more (32%) used government webpages, and 1.5 million more (72%) made online donations, amongst other behavioral changes (Americans were increasingly online anyway, meaning it is important to balance this broader context).[23]

In sum, the web and internet played a key role in the 11 September attacks. People used the networks in varied ways, from how people learned about the attacks to how they formed their thoughts and opinions, to how they connected with each other to discuss, and how they ensured that friends and family were safe. The online experience is important to histories of the attacks, of the internet, and of our collective response to the tragedy. How would memory institutions respond to capture and preserve this rich and important record?

"Whatever It Takes, Whatever It Costs, Do It": Archiving the Attacks

The attacks were chaotic for everybody. In Washington, DC, the Library of Congress's main campus was evacuated alongside much of the US federal government. Around the world, institutions including the Internet Archive as well as other web archives (for example, the first crawl of CNN.com archived at the Internet Archive that day comes from a donated collection from the French Institut national de l'audiovisuel) dramatically accelerated their crawls of relevant sites.[24] CNN, for example, was being harvested once every two or three weeks on average before the attacks. Starting at 4 P.M. Eastern Standard Time, CNN's website was harvested 45 *more* times before the day's end. It was collected an additional

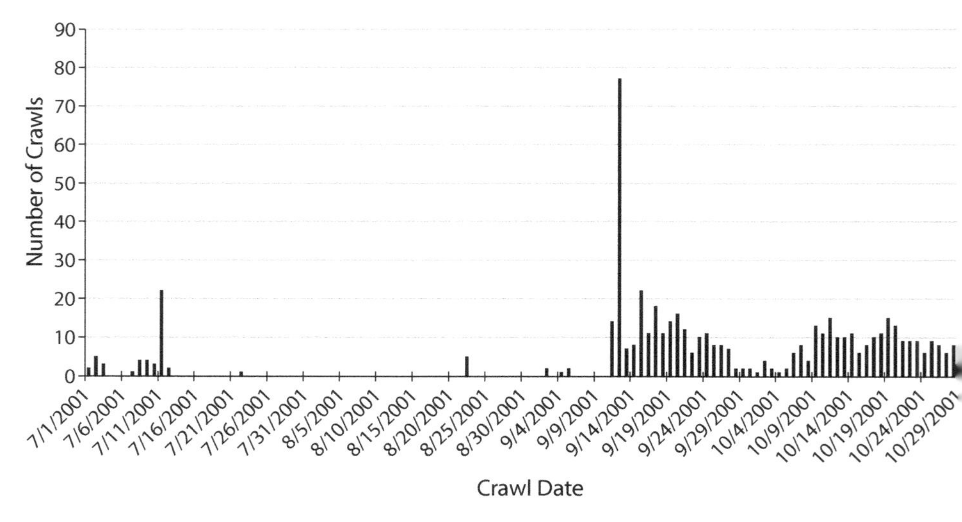

Figure 5.1 Crawl frequency of Nytimes.com. Data from Internet Archive, Wayback Machine, https://web.archive.org/web/20010801000000*/https://www.nytimes.com.

136 times on 12 September alone. The same is true for the *New York Times* website, as seen in figure 5.1. Such rapid moves ensured that rich documentation of the attacks was collected.[25]

In the day or two following the attacks, curators began to collect material. This was an exception to usual practice, which was to wait for "significant time and distance."[26] As it was then known, digital information was vulnerable. The rapidly growing online reaction to the attacks could prove to be ephemeral as Americans recovered from their initial shock. Enter the Library of Congress.

On the morning of 12 September, Library of Congress staff woke up, returned to downtown Washington, DC, and set to documenting the unfolding events. Cassy Ammen, a MINERVA team member, recalled that the Library of Congress's response quickly took shape: "When we returned to work on Wednesday, I received a call from Diane Kresh, then Director of Public Services Collections, with instructions to start archiving immediately. 'Whatever it takes, whatever it costs, do it,' she said."[27] Abbie Grotke remembers the collective memory within the Library of Congress as "we were all like, woah, we got to do this."[28]

MINERVA drew on existing connections with the Internet Archive and the site WebArchivist.org. The three organizations were collaborating in planning for collecting the 2002 American federal election.[29] WebArchivist.org was codirected by two researchers: University of Washington communications professor

Kirsten A. Foot and State University of New York Institute of Technology professor Steven M. Schneider. This would be a fortuitous partnership.

Schneider and Foot had created their own archive of the 2000 election to understand the internet's impact on it ("we were social researchers, not archivists," recalled Foot). In 2001, the two researchers learned that the Library of Congress had begun its web archiving pilot, and they pitched a proposal to collaborate with it for the 2002 election.[30] The Library of Congress was interested, and preparations began. In the run-up to the election, as part of this partnership, WebArchivist.org received additional funding from Pew Research Center's Internet & American Life Project. Alongside Pew and the Library of Congress, WebArchivist.org would carry out a project to archive K–12 school board sites as a "proof of concept and actually give us the chance to build out the [WebArchivist.org] interface before we got into the election."[31] That pilot began the first week of September 2001, inadvertently laying the foundation for the 11 September crawling to come. The school board project would soon be forgotten.

The school project, however, laid the groundwork to swing into action only two days after the attacks. The group began their work on 13 September. As Foot recalls: "That's why we're able to jump on it, because we've done all that pre-work for months to set up a project that was long term in vision because it's going to be about the 2000 election . . . [We'd planned] just a completely noncontroversial school digitization thing to make sure that everything works and that we write the code that we need to write to do what we really want to do . . . And that's what made that 9/11 archive possible in that regard."[32] The Library of Congress and WebArchivist.org would work together to select sites for crawling, while the Internet Archive collected the data itself. This collaboration made for a unique web archiving workflow as the time pressures and scale to be collected were significant.

Recommending officers from across the Library of Congress submitted sites from their subject domains to the web archiving team. This was a monumental task, involving colleagues across the entire library. As Grotke recalled: "Cassy [Ammen] was coordinating the internal efforts to identify [sites]. So she basically put out a call to every subject expert in the library . . . and it was very much like sending your URLs to Cassy, and then she was getting them in the crawl."[33] WebArchivist.org also crowdsourced user submissions on its webpage. The depth of experience and skill meant that the group scoped a significant collection. The Library of Congress would focus on "sites that spontaneously appeared as immediate reactions to the events—sites that were not easily found by search engines (too new to be indexed) but that were being shared in emails, linked from other sites, or in news articles at that time."[34] Ammen read the newspaper every evening for ideas on which sites to collect, which was then given to colleagues to find the exact URL to pass to the Internet Archive.[35] This labor paid off in the shape of an exhaustive number of nominated sites. Somewhere around 30,000 seeds were found. Between September and December, a subset of between 1,500 and 2,000 sites were collected.[36] Each site was assigned a subject term so that usable metadata could be assigned for cataloging.[37]

While the Library of Congress's earlier web archive had been a selective one, the 11 September archive aimed to document the experiences not only of Americans but also those around the world. It was a comprehensive approach to collecting, a significant collection strategy pivot. As Kresh wrote a year later, "the Library wanted everything—American and international reactions to the events of September 11, responses from the U.S. government and the military, as well as the responses of religious, ethnic, mental health and educational communities."[38] Project members crawled listservs and newsgroups. As with the earlier MINERVA project, harvested websites were notified. Sites that were included occasionally noted that their selection was an external endorsement of their significance. The Anti-Defamation League, for example, put out a press release explaining how honored it was that its "online response to Sept. 11 will become part of the national historic record."[39] Foot noted that the collection "really tried to capture how so many different kinds of organizations and entities and individuals were making use of the web as a structure for action."[40] As noted earlier, it would not be until 2002 that the Library of Congress moved to a stricter permissions-based process.[41]

WebArchivist.org reached out to the public to find more websites to collect. As part of its process of generating seed lists for the Internet Archive, Web Archivist.org built an "Archive This" button that could be incorporated into a web browser. "That tool sent an email to me," Schneider noted, "and generated a file and we sent the URL to the Library of Congress, and over time, we built the workflow to send URLs to the Internet Archive and they picked them up. And that's what spawned the archive."[42] It was through this approach that many "personal memorial sites," were archived, as well as discussion boards. This led to a different collection than what would have been found through the algorithmically driven crawlers. As Foot recalled: "Because what we are doing is a more proactive thing than what the Internet Archive had done all that time. [It] had just been following its browser paths and, you know, capturing whatever its spiders had touched. We were saying you got to go proactively and look specifically for these kinds of actors, these kinds of content, and then go from there."[43] Given the broad impact of the attacks, and the ephemeral nature of many of the sites, this was a fortuitous and forward-thinking decision.

WebArchivist.org's ever-evolving homepage throughout September 2001 and beyond lets us understand the collection's ever-evolving scope. By 17 September 2001, WebArchivist.org asked users to set up the "Archive This" bookmark in their web browser that would let them submit sites, or for more advanced users, they could also use an application called "NoteThis" to directly provide metadata.[44] While all content related to the attacks would be accepted, WebArchivist.org was "especially interested in finding sites by individuals—that record their feelings, experiences or opinions," as well as non-American ones.[45] Profiled in the *Chronicle of Higher Education* on 18 September 2001, Schneider stressed that "new things emerge so quickly that you have to archive it as it's being created or you simply lose it."[46] This close collaboration, involving crowdsourcing, traditional collecting, and technical collaboration, would quickly see a public portal established.

On 11 October 2001, september11.archive.org launched with over half a million accessible pages, all "related to the terrorist attacks and the United States reprisals, ranging from daily news reports to personal memorials."[47] It is worth underscoring how quickly that the site appeared: only a month after the attacks. Looking back, Schneider recalled all the work that went into getting the site live in such a short amount of time: "That was all Brewster Kahle. And his belief was like, 'we've got to go live!' " Conversely, Schneider recalled, the Library of Congress was far more hesitant.[48] Kahle explained that this was due to his worries that the "the collection will have holes in it . . . One of the reasons to get it out there quickly is so people can say, 'You're missing this.' "[49] Preservation and access were connected.

Drawing on the websites crawled by Internet Archive, as selected by WebArchivist.org and the Library of Congress, september11.archive.org allowed visitors to quickly navigate the vast array of half a million webpages, spread over some 40,000 sites. Metadata was written by Library of Congress experts. Pages were indexed, allowing for searching by date, keyword, URL, or webpage title. Some prominent sites received a subject heading, which allowed various pathways of exploration such as searching sites in terms of where Americans "advocated" (activism), "aided" (charities and donations), or sought information. Users could also explore government websites and businesses. A 13 October 2001 site screenshot is in figure 5.2.

As time went on, the september11.archive.org page evolved. By 2005, it encouraged visitors to instead visit the MINERVA collection at the Library of Congress. Then some 2,300 sites had been fully processed and cataloged by the Library of

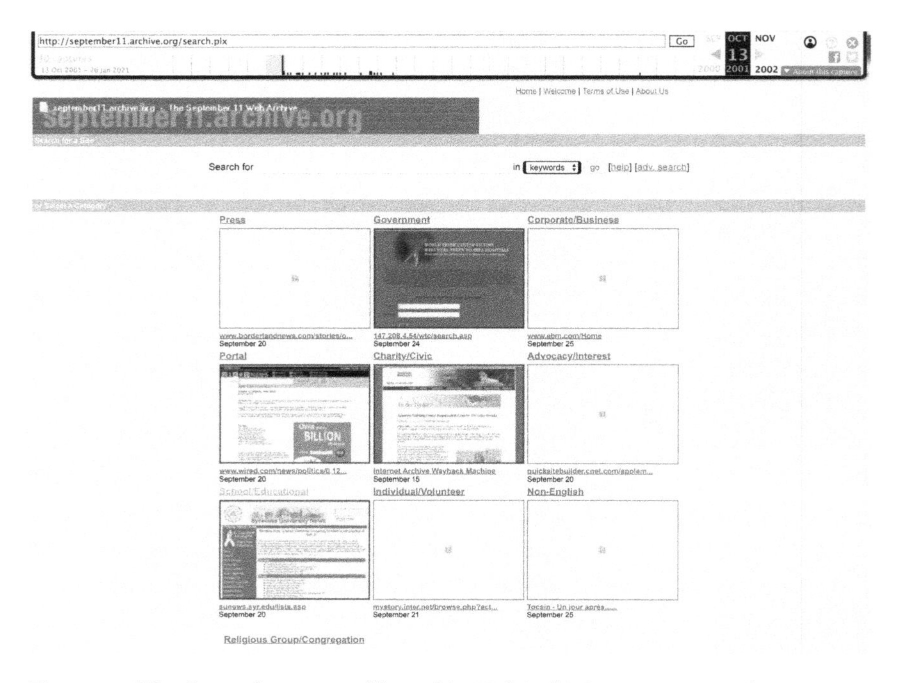

Figure 5.2 The September 11 portal hosted by WebArchivist.org on 13 October 2001. Missing images are an artifact of the Internet Archive's collection process. Courtesy of Internet Archive

Congress. There, MINERVA staff described their collection as preserving the "expressions of individuals, groups, the press and institutions in the United States and from around the world in the aftermath of the attacks in the United States on September 11, 2001."[50] With the later reorganization of the Library of Congress's web archive, the collection continues to exist, with some 2,200 websites accessible for replay and search. It vividly demonstrates the labor that goes into the creation of robust cataloging metadata.

For the Library of Congress, the response to the 11 September 2001 attacks was a watershed moment. The web archiving program was forever transformed. As Grotke recalls, the attacks were significant for the program: "We were a pilot thing and then 9/11 hit and we became a program . . . It legitimized what we were doing."[51] Thanks to the connections that the Library of Congress had with the Internet Archive and WebArchivist.org, it also set a precedent for collaboration. "9/11 was unique, and the Internet Archive could just start crawling," Grotke noted. "And I used to talk about the phone tree of urgent events. These events would come up, and we'd reach out to other web archiving colleagues and ask, are you crawling? Should we start crawling? Can we help in any way? . . . We [the Library of Congress] weren't ever good at crawling fast. But we had subject experts that could select stuff really well."[52]

Foot recalled the impact that WebArchivist.org and its collaboration had on the Library of Congress. She noted a conversation with a librarian there who acknowledged that before the collaboration the library's approach had been different: "Library of Congress had viewed web collections as like, 'here's a shoebox, put all of the stuff in the shoebox, and catalog the shoebox.' And we said as scholars, for anybody who's ever going to make sense of this, there has to be an accession record in library speak or metadata in academic speak about every element in that shoebox. And they have to be put in relationship to each other. And that was a paradigm shift for the library."[53] Indeed, the granularity of metadata helps scholars and others access web archives, rather than very broad notes at the collection ("shoebox") level. Usable collections require granular metadata, which requires considerable and concerted efforts.[54]

Web collecting would not be enough. The broad nature of the 11 September 2001 attacks necessitated the collecting of other forms of born-digital content: from email to web-based video games to digital voice recordings of people. In chapter 3, we saw how the Internet Archive had begun to expand into the world of a universal library: thinking beyond web pages to other forms of digitized content. The 11 September 2001 attacks would have similar broad impact on the heritage community.

The Universal Library Redux: The September 11 Digital Archive

A vibrant response to documenting the digital heritage of the attacks came from historians at George Mason University's Center for History and New Media (CHNM) and its partners at the American Social History Project / Center for Media and Learning (ASHP) at the City University of New York (CUNY) Graduate Center.

The roots of this response stemmed from another project. CHNM had just launched an innovative memory project: ECHO, Exploring and Collecting History Online. It had done so with a $720,000 grant from the New York City–based Alfred P. Sloan Foundation. Jesse Ausubel, a Sloan Program officer, had been funding projects around "digital collecting and thinking about the Internet as a medium for something that looks like oral history."[55] ECHO had the goal of recording the "the ever-expanding, ever-accelerating history of recent science and technology using a contemporary technology well suited to such a daunting yet critical task: the Internet." This project would do so through a combination of annotating web resources, helping historians of science and technology build their own websites to showcase material, and crucially, to let scientists and others in the STEM fields fill out forms to record their experiences in ECHO's "memory bank."[56] The site aimed to record memories on topics as varied as first memories of automobiles, planes, and television to the moon landing or vaccine experiments. ECHO aimed to create a comprehensive record of twentieth-century scientific life. The Memory Bank approach would become pertinent in the wake of the 11 September attacks, and the approach used to capture memories of the attacks.

For public historians realized that effort would be needed to preserve and document the attacks, complementing the Internet Archive and Library of Congress's work. Much of this record was digital. This included voicemails left from people trapped in the towers, leaving tearful goodbyes to their families and friends; email conversations between people on listservs, trying to make sense of the event, or to ensure each other's safety; to the sheer volume of flyers, photographs, and shrines placed around New York City. To many in the memory community, all these things—or at least much of them—needed to be documented, preserved, and cataloged.

On 4 October 2001, the Museum of the City of New York hosted "The Role of the History Museum in a Time of Crisis." Seventy public historians from thirty institutions came together. Attendees discussed both what should be collected as well as how to do it.[57] James B. Gardner of History Associations, who had co-organized Documenting the Digital Age, helped to organize this event. Out of the meeting emerged 911history.net, a hub for institutions to help coordinate "historians, museums, archives, and other collecting institutions" as they preserved and interpreted the attacks.[58] It was the follow-through missing from earlier events.

In November 2001, the Sloan Foundation approached CHNM as well as the ASHP. Sloan asked them if they could together create a "digital memory" bank of the 11 September attacks, similar to ECHO. As Stephen Brier, a CUNY historian and ASHP director, recalled, Sloan was concerned that "without a coherent and deliberate plan to capture digital materials related to the 9/11 attacks such materials would be lost." They were, to Brier, "especially focused early on in wanting to assure the preservation of stories of individuals who had personally experienced the events of 9/11."[59] CHNM would do most of the back-end database work as well as host the server, while ASHP would be responsible for the front end. Outreach was also divided. ASHP would do focused community-based outreach, especially in and around New York City, while CHNM would carry out broad outreach across the internet and local outreach in the Washington, DC, area.[60]

In early March 2002 the project secured funding. The two organizations received $700,000 to support the "September 11 Digital Archive project." Historian Tom Scheinfeldt was hired as a postdoc to be CHNM project lead. George Mason University's provost, Peter Stearns, a historian and editor of the *Journal of Social History*, underscored the project's significance. While he spoke clearly as both a historian and administrator, this was a vote of confidence in the value of preserving ephemeral information—something that had been largely unremarked upon by historians only half a decade previously. "There is a tremendous risk that a substantial amount of this information will be lost," Stearns noted, adding that even "after the passage of only a few months, the amount of social and cultural information we've lost is significant—not just the emotions and experiences of that day, but those beyond the specific tragedy itself."[61] Social history would form the project's intellectual foundation. As Scheinfeldt later explained, most "historical accounts of major events . . . are written using government documents and official accounts. Relatively few have many first-hand, grass-roots versions of the events."[62] This would be a test of a new kind of record, complementing the work being done by the Internet Archive, Library of Congress, and WebArchivist.org.

The September 11 Digital Archive soft launched in December 2001, even before securing funding, with initial content seeded by the project team. This was followed by an official launch on 11 March 2002, the attacks' six-month anniversary.[63] As early as mid-January, the site encouraged users to tell their stories. An early feature story was that of the "Hashers," a group of runners who "mark[ed] their routes with a trail of flour (while drinking heavily)," and who thus (perhaps understandably!) ran into trouble after the post–11 September Anthrax poisoning crisis.[64] The site also featured a growing collection of emails, setting the stage to encourage others to deposit their digital records as well. By mid-2002, the Digital Archive received its first major accession: the sept11info mailing list. That list had been established on the late morning of 11 September for people to share their stories, experiences, and questions on the day and weeks that followed. Some 3,536 messages by over 1,100 distinct users were donated.[65]

The September 11 Digital Archive had a broad mandate from its inception. When users went to tell their story, a form asked them to "tell us about what you did, saw, or heard on September 11th. Feel free to write as much or as little as you like. After submitting your contribution you will receive a confirming email that will include a copy of your story for your records." Users needed only to provide their state and city, their story, a name, and email address, and to indicate whether the story could be publicly shared. Optional fields included zip code, age, gender, race, and occupation.[66] There was also an image uploader that would later be creatively used by submitters.

Many of the contributions would involve what scholar Courtney Rivard has termed "indirect experiences."[67] These were stories of people watching the attacks on television, blogging about it in the days that followed, or recounting where they were on the day itself. In this, the memory bank approach was similar to other platforms such as wherewereyou.org which asked users to explore "Where Were You When America Was Attacked?"[68] Garnering submissions required active encouragement, including posting on message boards, exchanging links with memorial sites, and pursuing broader partnerships. This was complemented by the most effective strategy: spreading the word through traditional media. As Scheinfeldt recalled, "that kind of outreach was probably the most important thing still at the time."[69]

As Rivard has argued, these indirect contributions "reveal the ways in which September 11th was positioned as national tragedy where those indirectly involved were nonetheless interpellated to feel a part of the tragedy by directly identifying with the 'Americanness' of the victims."[70] Direct experiences were certainly in the archive, but the majority of the September 11 Digital Library consisted of indirect experiences. In the wake of the exhaustive interviews that would later be carried out by the 9/11 Commission, which focused on direct witness testimony, these indirect contributions help make the site unique and special. It is a social historian's dream. Reflecting on this, Scheinfeldt told me that he saw this as one of the enduring impacts of the archive: "the experience of sharing information by email and seeing information online and reading information online, ultimately contributing information online [to the archive], kind of reinforced 9/11 as an event in social history as much as it was in political history."[71]

Users themselves pushed at the constraints of the site. Scheinfeldt recalls being surprised at the "unexpected diversity of materials" received. Part of this was due to the back end allowing people to upload things other than images via the image uploader:

> We had three [database] tables. We had one for images, we had one for emails, and we had one for stories . . . And it turned out that people started using that image upload function to upload all kinds of things that weren't images. You know, in some ways it was kind of a happy accident that we were amateurish enough not to restrict the upload by file type on the back end. Because otherwise people would have gotten error messages and we wouldn't have gotten all this stuff. So, people

> were there, uploading flash movies. They were uploading Microsoft Word documents. They were uploading PDFs, all kinds of PSD files . . . Really, anything you could create on a PC circa 2000 we ended up collecting.[72]

Within this diversity of experiences were a mixture of direct and indirect experiences, helping construct an expansive historical record. From Flash games to personal stories, this was a massive body of stewarded cultural information.

Questions of authenticity and copyright loomed over the collection. Major organizations saw their own efforts falter on these problems, due to policy issues as well as overall institutional risk averseness. The September 11 Digital Archive, as a smaller grant-funded project, could move forward where other organizations would have hesitated. This was reminiscent of the way the Internet Archive operated. With protection under the takedown provision of the Digital Millennium Copyright Act, the archive primarily relied on contributor self-declarations. "For the most part, we weren't too worried about it," noted Scheinfeldt, rather they "were more concerned with the urgency of the moment and collecting what we could while we could. And we thought we would deal with issues of copyright and authenticity later."[73] This echoed similar decisions taken by the Swedish National Library and the Internet Archive—an awareness that the ephemerality of digital content required quick action. Thorny issues could be sorted out later.

The September 11 Digital Archive rapidly grew, thanks to donations such as the sept11info listserv as well as thousands of individual ones. By September 2002, archiving efforts received renewed public attention. A year since the attacks, the Archive had over 10,000 items, had annotated and assigned metadata to 5,000, and was looking forward to accessioning another 25,000. The growth of content, as well as high-profile media attention from venues such as CNN, flooded the site with visitors.[74] This attention also meant that the page soon appeared on the first page of search engine results for the term "9/11."[75] Success begat success. Reflecting on this, Scheinfeldt recalled the server load in the wake of the CNN interview and other outreach events around the first anniversary of the attacks. The server was stored in a closet at CHNM. Scheinfeldt remembers some 800 simultaneous connections taxing the server's cooling ability. His colleague Dan Cohen joined him in springing into action as the server overheated. They were "standing there with manila file folders, fanning the server to make sure it didn't overheat!"[76]

The Digital Archive partnered with the Smithsonian's National Museum of American History. The Smithsonian's exhibit, "Bearing Witness to History," opened on the first anniversary of the attacks. The exhibit was framed around fifty physical objects starkly displayed on tables without cases or interpretation. "The intent was to give visitors an intimate experience," explained the museum's website, "that will help make this historic day more real in their memories and stimulate them to reflect on its significance."[77] The museum sought to add more stories to complement the displayed narratives: visitors were encouraged to write or speak about memories, subsequently to be added to the September 11 Digital Archive.[78]

Both the Library of Congress and the National Museum of American History also relied upon a partnership with the Digital Archive to gather web users' thoughts on the attacks. As with the web archive, this was necessary as the institutions struggled with problems of authenticity, provenance, and copyright. The solution: "What they did instead was they offloaded it to us. In both cases, what we did was we basically reskinned our collecting portal in the style of either the Smithsonian or the Library of Congress website and linked to it from their website . . . I mean, there was a notice saying, you know, you're leaving the Smithsonian, but the interface was consistent . . . We were building an online collecting portal for these institutions without them having any actual ownership or control of the materials."[79] Submissions coming through the institution-specific portals were tagged with the institution, meaning that the portals provided access only to the items collected by them. Yet everything was added to the main digital archive as well.

It is worth underscoring the flexibility of the September 11 Digital Archive project, especially when compared to traditional heritage organizations. Scheinfeldt was self-deprecating in our interview—"we were just a bunch of basically grad students who could kind of ignore" dimensions of collecting policies, copyright, authenticity, and procedures—but the relationship was broadly similar to that of the Internet Archive vis-à-vis the Library of Congress.[80] Without this flexibility toward collections policy and finding material on the web, ephemeral content and memories of that day would have been lost. It was not coincidence that one of the project's mantras was "perfect is the enemy of the good."[81]

The open collection policy meant that the digital record contained a diversity of political responses to the attacks. These ranged from mainstream to fringe, from unifying to racist. In the case of the digital archive, each submission was reviewed by a graduate student. In general, the project took a "light touch" and rarely censored submissions.[82] Ultimately, the site would contain material across the political spectrum from pro- to anti-Americanism.

All this content serves as an important counterbalance to official narratives. The official history was and is naturally politicized. The National Museum of American History became the "official" museum of the attacks. Congress passed a law to "collect and preserve in the National Museum of American History artifacts relating to the September 11th attacks of the World Trade Center and the Pentagon," the first official direction from Congress to the Smithsonian's collection in the museum's history.[83] While operational direction was left to the museum, the memory of the previous Enola Gay controversy—which had highlighted the impact of nuclear weapons on the citizens of Japan rather than uncritically celebrating the aircrew—would lead, as Rivard has convincingly argued, "to a collection that effectively functions as a national memorial."[84]

The same could not be said for the September 11 Digital Archive. While the Digital Archive serves a memorial role with narratives of loss, trauma, and redemption, it also contains counternarratives. These include revenge fantasies as well as anti-Americanism and xenophobia, both of which would be out of place in a solemn official history. Yet these are important components of the American public response to the 11 September attacks.

Rivard has discussed the "Collected Digital Animations and Creations" section of the Digital Archive in her work, noting that racism, sexual violence fantasies, and other "spectacular and horrific images of sexualized violence," represent "one of the few, if not the only, places that the digital 'mechanics of war' are actually preserved." Animated American flag GIFs may have been uplifting, but there was also rabid and racist anti-Taliban and anti–Osama Bin Laden Flash games to contend with. As Rivard concludes, "[w]ithout this collection, the power that the internet has played in constructing ideas of good citizens, national belonging, and enemy-terrorists in relation to September 11th may have been lost in virtual space."[85] Here, the open upload feature of the "image" benefited the historical record. Scheinfeldt recalled receiving animated Flash games, many of which involved killing Osama Bin Laden.[86] Indeed, the inclusion of content that is outside the mainstream political consensus is a benefit of the Digital Archive. Such material, even if it makes for difficult viewing, is critical to understand the moment.

While there was political diversity, other forms of diversity were lacking. The team realized that given that most of their unsolicited contributions came from active web users, they skewed toward those who were "largely white and middle class."[87] In part, this reflected the problems in 2001 with getting people to submit information on the web, as Scheinfeldt recalled: "The Web, circa 2000, was still for most users a read-only medium right? . . . Most Web users used it as a kind of telephone book where you wouldn't actually publish content. You would browse content . . . And so the notion that you would go to a text box on some website and tell some anonymous person or people what you were doing and thinking and feeling was pretty foreign to people, right? Like that wasn't something you did as a matter of course."[88] This illustrates a limitation of the historical record they were assembling. Those who were comfortable with using web forms were in general more affluent and thus unrepresentative of the many people affected by the attacks.

To rectify this, the Digital Archive team sent out targeted emails, school visits, and in-person community engagement to diversify the collection. These groups were as varied as Muslim Americans as well as residents of Lower Manhattan's Chinatown neighborhood (only a few blocks away from the World Trade Center site).[89] Scheinfeldt recalls that it was around the one-year anniversary that "we started to see, like oh, like we don't have anything from Chinese language speaking communities, even though they are living in the shadows of the Twin Towers. Maybe we should do a little more outreach there."[90] While CHNM focused on technical dimensions, community outreach mainly fell to the ASHP team. To engage communities, a vast array of activities were carried out: a Spanish-language version of the site was created; staff were sent to Shanksville, Pennsylvania, to collect local reactions to the United 93 crash; they worked with CUNY's Middle East and Middle Eastern American Center to connect with the New York City Middle Eastern community; and they carried out Chinese-language interviews with Chinatown residents.[91] Thanks to these efforts, coupled with the site's high profile, by 2003 the Digital Archive had collected approximately 150,000 items (60,000 emails, 14,000 digital images, 45,000 personal narratives, 6,000 documents, and 4,500 audio/visual files). It had received 120 million hits and over two million users since its launch.[92] Soon the project needed to ensure that all of this material would be stewarded in perpetuity. As with all efforts to avert the digital dark age, the problem of perpetuity was the big one.

Simply Delaying the Digital Dark Age? The Problem of Sustainability

The question of sustainability looms large with all digital preservation projects. The Internet Archive and national libraries continue to thrive today. But what might happen when a project begins to sunset? By looking briefly at the experience of the September 11 Digital Archive, we see the degree to which digital stewardship requires sustained investment. Otherwise, one is just delaying the inevitable. Information gathered today that is lost tomorrow benefits nobody.

How could the Digital Archive ensure its own long-term survival? A complicated database-driven site does not lend itself easily to web archiving. Grant-funded projects often face difficulties in long-term preservation after the original funded period comes to an end. Here, the connections between nimble projects and large, long-lasting institutions would appear again. In November 2003, the Library of Congress announced that it would make its first major 11 September acquisition—

as well as one of its first major digital acquisitions—by accepting the September 11 Digital Archive to its collection.[93] Just as the Internet Archive would crawl web content to be later donated to the Library of Congress for preservation, so too would this project. This arrangement would leave ongoing operations of the site to CHNM and ASHP, with the Library of Congress handling long-term preservation. Planning would begin. The actual transfer was planned to happen in roughly ten years, when servers and data would be given to the Library of Congress.[94] The arrangement suggested a path for long-term digital stewardship.

On 10 September 2003, the agreement was announced and formally executed, the occasion marked by a day-long symposium at the Library of Congress: "September 11 as History: Collecting Today for Tomorrow." Diane Kresh, who had spurred the Library of Congress into action, took a moment to recognize the collective effort made by so many to establish the historical record. "Even in the midst of the initial chaos of the horrific events of September 11, 2001, the Library of Congress began collecting materials documenting the attacks," Kresh noted, adding that the "September 11 Digital Archive, with its vast content of firsthand accounts, will add to the broad range and diversity of materials already acquired by the Library of Congress that relate to the September 11 tragedy."[95]

Sustainability challenges reared their heads even after the formal 2003 agreement with the Library of Congress. The infrastructure became increasingly fragile and unstable, especially as large grants inevitably began to dry up. As Brier and Brown noted, the project "did not end with collecting." Tasks involved infrastructure stabilization and security, metadata updating to make digital searching more successful (a perennial problem with user-submitted metadata is that it does not always line up: for example, "WTC" and "World Trade Center" might refer to the same concept, but absent their connection, their associated objects will be siloed), to build a better back end using open-source tools that CHNM was developing, and to redesign the website.[96]

In February 2011, the Digital Archive received a "Saving America's Treasures" grant from the National Park Service and the National Trust for Historical Preservation. Normally focused on buildings or physical objects (such as reactors or furnaces), this grant aimed to ensure the Digital Archive's sustainability. As the team explained when they received the award, "Cutting edge at its launch nearly ten years ago, the Archive now is showing its age."[97] Sharon Leon, a historian at CHNM and then its director of public projects, put it prosaically: "looking back 10 years, the site is held together with chewing gum . . . That's the fate of all the early projects [especially those that] were built without a sense of standards."[98] The Saving America's Treasures grant allowed the Archive to fully adopt its open-source technology platform developed at CHNM, Omeka, which would also serve as a foundation for other future memory banks. It would also tackle the list of sustainability problems identified above.[99]

The September 11 Digital Archive collected until October 2021, when after twenty years the team noted that "we feel that it is appropriate to transition this from an active collecting project to an archive of materials for study and reflection."[100] Even after this closure, the Digital Archive remains a vibrant site awaiting final archiving—despite the fanfare with which the Library of Congress accepted the hard drives, access remains only possible through the original site.

The 11 September attacks, and the September 11 Digital Archive and related efforts, were a model for what would later become a broader approach to "disaster collection." A subsequent 2005 project, the Hurricane Digital Memory Bank, built on the September 11 Digital Archive's foundation. Established in the immediate wake of the devastating 2005 hurricanes in the southern United States—Katrina in August 2005 followed by Rita in September 2005—CHNM partnered with the University of New Orleans to launch this new site. Sheila A. Brennan and T. Mills Kelly, CHNM professors who set up the Hurricane Digital Memory Bank, noted that they had created "a model of a participatory 'archive of the future'" with the September 11 Digital Archive, and they looked forward to similar success.[101] Funded with $200,000 from Sloan, the Hurricane Digital Memory Bank collected "first-hand accounts, on-scene images, blog postings, podcasts, and videos," via both its website as well as—cognizant of infrastructure damage—via a telephone number that would record experiences.[102] Unfortunately, the site did not catch on as had the earlier 11 September effort. Potential reasons included a clunky new user interface that may have lagged expectations of an audience accustomed to Web 2.0, as well as a focus on the direct rather than the indirect experiences that made up the September 11 Digital Archive.[103]

While later projects would not achieve the salience of the September 11 Digital Archive—no event in the United States has yet led to a similar sense of history unfolding as the attacks did, perhaps until the January 2021 Capitol attack—a new model of digital collecting had been established. Scheinfeldt reflected on the long-term impacts forged by the project:

> I think it did spark a recognition among large national collecting institutions that this was something that needed to be done. I think they knew this stuff was out there, they knew it would be important. But I think, you know, as I said, they couldn't kind of get out of their own way. And I think they realized that the next time something big happens, we're going to have to get out of . . . We're going to

> have to figure this out for next time, you know, because we can't just rely on grad students to keep doing this.[104]

This has been borne out. I wrote much of this chapter during the COVID-19 pandemic, where memory institutions around the world are furiously collecting the born-digital heritage of the events surrounding us. In 2008, with an eye to helping others establish similar projects, the CHNM and the Minnesota Historical Society released their Omeka software, an open-source platform that allowed other institutions to set up digital archives of their own. Some of the earliest sites created with this platform continued the tradition of event-based harvesting, such as a site to remember the tragic Virginia Tech shootings in April 2007.[105] Future event-based collections would include the 2012 Boston Marathon bombing ("Our Marathon"), Hurricane Harvey in August 2017, and the global coronavirus pandemic. Similarly, in the wake of the 2022 Russian invasion of Ukraine, volunteers and professionals collaborated to safeguard Ukrainian digital heritage and websites.[106]

When an event happens, new scholars and volunteers move into action. Citizens are empowered to create comprehensive digital collections of elections, viruses, turmoil, sports, cultural events, and beyond. It is no longer just national libraries or the Internet Archive that build collections, but all of us. Events like those of 11 September will be, thanks to the efforts of memory professionals, among the best-documented in human history.

Conclusion

Historians are beginning to write the history of the September 2001 attacks. Most focus on the attacks through the eyes of *direct* participants and observers: those on planes, in touch with loved ones through cellphones and airphones; first responders; air traffic controllers who tried to make sense of confusing information; or people in New York City and near the Pentagon.[107] Yet today the memories are reinforced by the testimonies of indirect participants. Even at the National September 11 Memorial & Museum, visitors record their memories even if they had not been in New York City during the attacks, or not even born yet in 2001.[108] Letting anybody join the conversation about the event contributes to a framing that sees "9/11 less as a specific event or date circumscribed by geographical spaces and more as an unending process of subjectification," reducing the privilege of direct testimony—perhaps even supplanting it.[109] The digital archives are part of this current.

One day, a comprehensive cultural history of the 11 September attacks and their aftermath will be written. It will explore the experiences of indirect participants across the country: collective thoughts, the emergence of narratives, and how these voices combined to create the cultural milieu that so many experienced during the attacks and their aftermath. This book has yet to be written, however, in part for technical and practical reasons. The metadata schema within web archives remains challenging, and access in general to the terabytes of data within these archives is difficult. It is hard to imagine how such a history, however, would be written without the platforms and collections discussed in this chapter. One impact of this collecting is clear: a robust documentary record from thousands of people awaits. Only the most ardent digital hoarder would wish that they had more, not less, information about these terrible events.

For the impact of the September 11 Digital Archive and the web archives is felt in that they are themselves part of the event's cultural history. The rush to collect memories from as many people as possible is part of the enduring legacy and construction of the event. Collecting voices from direct and indirect participants was in part a legacy of the capacities and approaches built up by web archivists, digital preservationists, and early digital historians. After 2001, the progress made around averting a digital dark age was clear in the concerted efforts of people around preserving events like these.

Today, when a major event unfolds, historians, librarians, and archivists quickly arrive on the virtual scene. In the wake of the 2015 and 2016 terrorist attacks in Paris and Nice, the Bibliothèque nationale de France and the French Institut national de l'audiovisuel were able to spring into action almost immediately, allowing scholars to understand reactions from people online almost immediately after an event happens.[110] Information is still lost or missed as there is no magic bullet to preserve everything, but many events witness a gap of only minutes between event and collection. Elections, protests, sporting events, pandemics—all see an unprecedented degree of information preserved, not just by professional librarians but by everyday citizen curators, academics, and activists. The selectivity of these memory activists and experts complements the broader ongoing collecting of national libraries and the Internet Archive.

The 11 September 2001 attacks were a watershed moment in digital preservation and collecting. The digital dark age, which had preoccupied so many over the previous decade, would be measured for much of the web on that day not in weeks and months but in hours and days. People turned to the web for information, to memorialize, to console each other, to post flags and ribbons, and in so doing, contributed to a rich and evolving historic record. When historians write the social and cultural history of that day and the weeks, months, and even years of official memory and memorialization that would follow, these sources will be critical windows into how people understood the attacks.

Our story of event-based capturing continues in our unsettled world. More recently, the COVID-19 pandemic has challenged us to collectively ensure our historic record makes sense of this seemingly singular moment.[111] Historians point to the earlier experience of the century-ago Spanish Flu pandemic to inform conversations, but they lament that much of that event was overlooked and that the historic record is smaller than we wish it was. To leave a more robust record for those who come after us, archivists and libraries around the world moved to act as the pandemic reshaped the world, documenting community responses, institutional actions, social media hashtags, and beyond. Global institutions, coordinated through the International Internet Preservation Consortium, worked together to collectively gather tens of millions of documents and over ten thousand archived websites.

Ultimately, the experience of documenting the 11 September attacks also underscores the dramatic transformations of the historic record that had taken place over the preceding decade. Ideas and platforms to capture digital information had been developed, and only hours after the attacks, they were able to swing into action. Archives shape the histories that we write. As the 11 September attacks were a watershed moment, it seems clear that the digital archives and records have led to a much more substantial and inclusive historic record. This will be one of the enduring legacies of our averting the digital dark age.

Conclusion

Constantly Averting the Digital Dark Age

The digital dark age was largely averted by 2001, at least in terms of the most apocalyptic predictions made only a half-decade earlier. The widespread destruction and loss of digital heritage would not happen. With the establishment of the Internet Archive and national library programs, the big question was how sustainable their holdings would be over the long term. This was like the dilemma facing the September 11 Digital Archive. As the web and its archives aged, would its memory endure? Without active investment, digital content disappears. Furthermore, could collecting institutions continue to innovate and rise to the challenge of new trends and technology? The web constantly changes, from new platforms to dynamically generated content. A 2002 web crawler could not capture the dynamic web of 2012, let alone 2022 or today. Perpetuity is a long time. Perhaps it is hubris to imagine that the digital dark age has been averted? Perhaps traditional archiving is an impossibility in the "age of algorithms," with personalized experiences instead requiring that we document user experiences.[1]

Fortunately, the field continues to rapidly evolve. Institutions have not rested on their laurels. In 2002, cognizant of the value of having more copies around the world to steward the long-term health of the Internet Archive, as well as given its symbolic geographical value, the new Bibliotheca Alexandrina in Egypt acquired a copy of the Internet Archive's holdings.[2] The first Library of Alexandria had aimed to be a universal library. How better to help create the new one than by donating a comprehensive web archive? National library web archiving continued to expand around the world as the technology became more accessible and commonplace. While still a field dominated by the Global North, it is slowly expanding beyond its Western roots. In 2004, the National Library of Korea began the Online Archiving & Searching Internet Sources (OASIS) project, creating its own national web archive. Then, in 2010, the National Diet Library of

Japan was given the authority to archive public agency websites, and in 2019 the National Library Board of Singapore began comprehensively crawling its web domain as well.[3] In the aftermath of the 2016 American federal election, as Donald J. Trump assumed the presidency, web archivists and allied activists around the world acted to ensure the preservation of vital American government data. Much of the focus was on climate change data, but the importance of documenting the transition became global news.[4] Other events, from the global pandemic to wars to subsequent terrorist attacks and protests against racism and police brutality, continue to underscore the need for web collecting. Eternal vigilance is key. Ephemeral content shapes our world. Collecting is a constant task, carried out by vigilant professionals.

The Digital Dark Age Never Dies

Even in this, the age of widespread preservation, the idea of a digital dark age continues to haunt us. It is a concept with surprising longevity. Perhaps this is because the phrase takes a complicated problem and evocatively summarizes it. In 2003, a syndicated wire service story again raised the fear that even if an adult reader could look at traditional, yellowing photos of their childhood, their "grandkids probably won't fare as well with your digital photos. The computer files may survive but the equipment to make sense of them might not. This era could become a 'digital dark age,' a part of its collective memories forever lost."[5] This framing now ignored the work that went into these problems. Tropes and themes from the mid-1990s are continually reborn. The digital dark age endures.

This is frustrating. Save apocalyptic collapse, the digital dark age has been averted. Anything that would lead to a digital dark age would presumably lead to a nondigital dark age as well. In such a case, perhaps we would instead be looking to *A Canticle for Leibowitz* for inspiration. The stewardship of digital documents is a solved problem insofar as we understand the need to shepherd information (not to say that new formats, platforms, or copyright restrictions do not cause trouble). If the Library of Congress ceases to exist, the digital information it holds may disappear as well—but so would its other holdings. We should not discount the need for an organization like the Long Now Foundation, but we also should not irrationally fear that our collective digital heritage can disappear in the blink of an eye.

Fears of the digital dark age continue unabated, seemingly disconnected from the developing digital preservation field. In 2003, a letter published by *Nature* raised the fears of a digital dark age of email, worrying about the "potential for the loss of records that may have immense historical value."[6] In 2009, the metaphor shifted somewhat with chilly fears of a "Digital Ice Age" ("we may find our files frozen in forgotten formats").[7] The next year saw Kurt D. Bollacker advising in *American Scientist* around how society could avoid a digital dark age so that we could ensure the preservation of "the record that future generations might use to remember and understand us."[8] Indeed, well over a decade since the original coining of the term, archeologist Stuart Jeffrey raised the specter of a "new" Digital Dark Age in 2012, as his colleagues began to use commercial tools in an "open, dynamic, and fluid" context that perhaps eluded traditional preservation approaches.[9] Jeffrey's point was an important one. Digital preservation requires continual engagement and investment—which it is indeed seeing.

Most vividly, in 2015, Vint Cerf—then a Google vice president—declared at the American Association for the Advancement of Science annual meeting that without action to avert digital loss humanity faced a "forgotten generation, or even a forgotten century."[10] Cerf's intervention that year probably gave the idea of a digital dark age more attention than it had ever had. Stories on the topic appeared in the *Atlantic*, the BBC, and media outlets around the world. Scholars and practitioners in the field were disappointed. Michael Nelson, a computer science professor at Old Dominion University and a leading web archives researcher, likened the media frenzy around Cerf and the digital dark age as akin to "having your favorite uncle forget your birthday, mostly because Cerf's talk seemed to ignore the last 20 or so years of work in preservation."[11] Of course, Cerf had been part of the first generation of agitators involved in raising the need for web archiving and digital preservation more generally. Yet even if Cerf had overlooked some work that had been done, perhaps it suggested that the memory community needed to better articulate its value and activities. In some ways, it was more a problem of publicity and awareness than that of preservation. It also speaks to the lack of a public understanding that the web remembers, which has profound consequences for those who unwittingly forget about institutions such as the Internet Archive. Many a politician would have avoided controversy had they known that the Internet Archive existed.

While Cerf became increasingly involved with the International Internet Preservation Consortium and modern web preservation, the idea of a digital dark age persisted. Another Google employee, Rick West, raised fears of a digital dark age in 2018, in terms that echoed the prospects raised almost fifteen years earlier. West noted that society "may [one day] know less about the early 21st century than we do about the early 20th century . . . The early 20th century is still largely based on things like paper and film formats that are still accessible to a large extent; whereas, much of what we're doing now—the things we're putting into the cloud, our digital content—is born digital."[12] The persistence of these tropes suggests to some degree an appetite from the worried public or journalists to explore information's ephemerality. It may also reflect the growing role of digital loss in our own lives. Many of the earliest digital preservationists were spurred into action by their own experiences. How many of us now face digital instability in our daily lives? Attention is not bad. By worrying about a digital dark age, we avoid it.

Some of this anxiety is well placed, especially when specifically aimed at the problem of preserving material created or posted on third-party platforms such as Facebook and Instagram. As private user-generated information is locked behind passwords and beyond the reach of web archivists, the data people put on Facebook is at Facebook's mercy. This raises genuine concerns around a renewed digital dark age: not file formats or lost websites per se, but rather how our memories are deposited on private platforms. As Adam Shepherd evocatively asked in 2019, "How many of the photos and videos that you've shared on Facebook and Instagram do you have copies of in other places?"[13] My personal website will be preserved by the Internet Archive, and our national library in Canada will have copies of news sites and government pages that provide broader context to my life, but most of my personal photos and thoughts live on Google Photos and Twitter (now, sigh, known as X). If those sites were to shut down (and one day they will), would my earliest photos and tweets about my children disappear too?

The specter of a digital dark age thus forces us to continually think about the longevity of digital information. If your Facebook account was hacked, or if you died, or if you wanted to ensure somebody had access to it in the medium-term future, would your information be safe or accessible? If not, what steps could you take to do so if you wanted it to be?

Will web archiving become a victim of its own success? The institutionalization and routinization of web archives runs the risk that they are taken for granted. Seemingly reliable operational initiatives do not always get the attention that they deserve. While web archiving today is a core library function, for most institutions it is minimally staffed. Web archiving is done on the side of desks, with few dedicated personnel. In the United States, for example, in 2017 four out of every five web archiving institutions employed less than one person to do this task.[14] This is a lot of critical work falling on few shoulders. The ease of web archiving has not been matched by investments to improve the overall capacity of web archiving.

The COVID-19 pandemic brought the implications of this underresourcing into relief. In the first few weeks of the crisis, web archivists moved into action, documenting spreading lockdowns across much of the Western world, preserving images of empty streets, and social media conversations that debated public health interventions. Yet as weeks turned into months, archivists began to feel what was described as "curatorial fatigue."[15] Without sufficient support, individuals were overwhelmed by the importance of the task layered on top of other work responsibilities and personal challenges. In the wake of the 11 September 2001 terrorist attacks, teams of curators worked together to document events and establish portals. There was a spirit of innovation and of breaking new ground. Now, with web archiving an established yet peripheral function, too many librarians and archivists found themselves overwhelmed. In other words, a tall order has been set for many of our web archivists today: no less than deciding the fate of much of our historical record. A lot of weight is put on their shoulders. Algorithms alone will not avert the digital dark age. People will.

The Right to Be Forgotten and the Pitfalls of Averting the Digital Dark Age

When the Internet Archive first became known, many observers were alarmed. What about the privacy implications? In July 1996, David Berreby encouraged his *Slate* readers to consider the "most embarrassing e-mail you ever wrote, available to anyone curious enough to go looking . . . As we're encouraged to exult over the vast new volumes of information that are becoming easier and easier to capture, remember that the art of losing is also important to master."[16] Dan Gillmor similarly worried in September 1996 about the prospect of "every dumb thing [that] I've said on-line" being saved.[17] A year later in March 1997, John Markoff worried that the bigger the web archive, the bigger the Big Brother problem would be.[18] When the Wayback Machine launched in 2001, it provided relatively complete access to the Internet Archive collection. However, as a user needed a URL until the recent advent of limited keyword search, the Wayback Machine provided privacy through obscurity. The apocalyptic visions of 1996 did not come to pass. Both public and private people alike have been affected by the Internet Archive, and things may have been remembered that they wish were not.

Yet by 2010, there was also increasing concern about the impact of old, decontextualized information. Viktor Mayer-Schönberger's 2009 *Delete* traced in part the collision between Web 2.0 (by 2001, as he notes, "users began realizing that the Internet wasn't just a network to *receive* information, but one where you could *produce* and *share* information with your peers")[19] and the accessibility of web-based personal information on the web. At times Mayer-Schönberger overstates his points—the argument that "forgetting has become costly and difficult, while remembering is inexpensive and easy" is not true.[20] However, his argument that "comprehensive digital memory represents an even more pernicious version of the digital panopticon. As much of what we say and do is stored and accessible through digital memory, our words and deeds may be judged not only by our present peers, but also by all our future ones" is profound.[21] Reactions to this emerging trend have taken different shapes around the world.

The "right to be forgotten," as a legal precedent, stemmed in part from a Court of Justice of the European Union ruling in the 2014 Google Spain decision. The right drew on deeper histories, concepts, and precedents. The case involved a Spanish man who was unhappy that a Google search for his name surfaced details of a government auction of his property. He felt this resolved situation was no longer relevant and impugned his reputation.[22] The Court of Justice ruled in the man's favor, holding that Google had a responsibility to balance its rights with those of users, giving the plaintiff the right of erasure.

This "right to be forgotten" influenced the European Union's subsequent 2018 General Data Protection Regulation. GDPR enshrined the right of individuals to request that their data be erased for specific reasons (which then needed to be balanced against free speech rights). As of 2020, approximately 45,000 requests were lodged with regulators to have information delisted from search engines in Europe, of which just under half (43%) were deemed to be valid.[23] Reflecting different legal and cultural traditions, many North American commentators reacted with surprise to these decisions and regulations. Would the "right to be forgotten" not limit freedom of expression?[24] The leaders of the Wikimedia Foundation (which oversees Wikipedia) worried about these rulings creating the prospect of "an internet riddled with memory holes—places where inconvenient information simply disappears."[25] Would all the hard work of web archiving be undone if inconvenient documents could disappear through legal fiat?

Legal scholars Melanie Dulong de Rosnay and Andres Guadamuz argued that ultimately the legal requirements would have little to no impact on web archives, given the already existing policies and approaches to privacy.[26] The Oakland Archive Policy already implicitly gave most people a right to be forgotten. Conversely, as much of the right to be forgotten involves delisting content, the combination of the lack of full-text search in most web archives as well as the relegation of national library legal deposit collections to on-site only access had already established a balance between privacy and access. As one does not need to delete content to conform with the right, only to make it less accessible, web archives were compliant.

Improving computational access may unsettle this balance. Some of the original ideas around access, such as treating archives as akin to census data, had tried to strike a balance between privacy and access. Since the Wayback Machine's launch, most models have implicitly adopted a privacy-by-obscurity approach. In the case of national libraries, some have been required to restrict access to on-site reading rooms. On-site access, which often strikes observers as ludicrous—forcing a researcher to physically travel to sit at a computer to view networked resources—is a reasonable compromise between access and privacy. Sitting in a reading room on a special terminal, few readers would mistake archived webpages for live content. Context collapse can be avoided. Parallels can be drawn with the digitization of nondigital archival sources, which also raise ethical access questions.[27] Giving access to researchers in a reading room *is* different from unfettered decontextualized access to anybody with a Google search query, even if both collections are, in theory, open. Conscious of the burden that travel places on many researchers, particularly those with limited travel funding or caregiver responsibilities, perhaps a similar outcome could be achieved through a Virtual Private Network (VPN). Some degree of friction is not a bad thing.

One fear of unfettered digital access is the prospect of an archived website or social media post being consumed out of context and used to shame somebody. For example, in my earlier archival work on 1960s student activists, I encountered documents that raised questions about how my reaction might have been different had these activists tweeted or even had all of their material digitized rather than having letters that ended up in archival boxes.[28] One letter, from the McMaster University Archives and Special Collections, contained 1962 musings by a student activist in their late 20s who thought it would be "fantastic news" if nuclear-tipped Bomarc missiles (which were to be used over Canada in an anti-aircraft defense strategy) could be redirected by the Soviet Union after launch to turn around and destroy their launch bases. This letter helped me understand the context of the New Left, presenting a vivid example of how the New Left was very different from the mainstream Canadian social democratic left. Yet the thought process of that letter *in isolation* is horrifying on the surface. It effectively mused about the deaths of Canadian military personnel and those nearby. But as part of a broader collection of correspondence, it illustrated the ways in which New Leftist intellectual thought was developing.

As a researcher sitting in an archive, I found the letter to be unremarkable. Yet if a keyword search for the author would surface it, readers might be less charitable sans context. Suddenly the need to travel to another city to visit an archive or log into a cumbersome VPN is less of an inconvenience but perhaps part of thoughtful, historical research. What *if* that letter had been a tweet or a blog post in a web archive? A student writing that today could become the target of an outrage campaign.

In 2016, my colleague Nick Ruest and I were bullish on the prospects on what social media meant for future historical research. "Consider what the scale of this dataset means," we wrote. "Social and cultural historians will have access to the thoughts, behaviours, and activities of everyday people, the sorts of which are not generally preserved in the record."[29] I still stand behind this. Historians, professionally trained and responsible ones, conscientiously grappling with the past's complexity, will try to understand the proper context of archival documents. When I discovered the 1962 Bomarc missile letter, I did not turn to a conservative Canadian media outlet, claiming to possess the smoking gun of the New Left's moral degeneracy. Context matters: private correspondence, part of an ongoing debate around the adoption of nuclear weapons in Canada, by a young student exploring new ideas in a climate of intellectual exploration. I suspect few professional historians would think differently.

On balance, I believe the value of a society-wide "right to be remembered" outweighs *in general* the value of an individual's right to be forgotten. This needs to be considered in a context of archival ethics and care. The compromise position of complicated on-site or remote access, requiring researcher registration, compels researchers to view and think about documents in context. By doing so, readers know that web archives are not just websites like any other on the live web. Furthermore, researchers cannot easily share the "gotcha" moment, flattening time as if a blog post from 2001 is the equivalent of one in 2021.

Historical processes unfold by virtue of human choices and decisions. The current situation when it comes to digital access was not inevitable. The international library community has generally taken a conscious choice toward expansive selection. In the case of the Internet Archive, it ultimately adopted relatively open access policies, but it handled privacy through obscurity. As institutions develop next-generation search and retrieval systems, the lessons of the past serve as concrete inspiration about alternate visions of web archival access.

The Little Digital Dark Age, 1991–1996

The direst predictions of a digital dark age were averted by the beginning of active web preservation in 1996. Yet there was a short dark age after all, between the advent of the web in 1991 and the development of memory institutions in 1996. In some ways, this ultimately formed the dreaded "digital gap" that commentators such as Danny Hillis had worried about. Seeing what we have lost can help us gain a better appreciation of what has been saved.

Not everything published on the web before late 1996 has been lost. Magazine articles, newspaper accounts, oral interviews, journal articles all help to reconstruct the earliest web. Digital forensics can as well with considerable effort. Tim Berners-Lee's first website, for example, had been launched in December 1990 at CERN. By the time web archiving began in 1996, it had been converted into a museum site. An effort to reconstruct it drew not only on preserved files but also an array of other contemporary sources to recreate the browser experience as it might have been in 1990.[30] Similarly, the first American webpage, the 1991 homepage of the SLAC National Accelerator Laboratory was restored from backup. Researchers needed to undertake a fairly involved process to convert an "original list of scattered files into an accessible and browsable website."[31] Apart from these rare exceptions, however, most of the actual websites from this period are lost. The labor needed to reconstruct sites like this does not scale. We can reconstruct a few significant sites with great effort and the fortune of either backups or contemporary documentation. Most sites are beyond this.

What have we lost? We can look to events that took place in 1995 and early 1996 to understand the gap. Before 1995, few corporations or people were on the web. Accordingly, 1995 to 1996 was the earliest period when the web was part of many people's lives.

Political examples are perhaps the most obvious. Historians can look to significant events to then in turn see what remains. In October 1995, the Canadian province of Quebec held a referendum on whether the province should pursue sovereignty. Digital historian Ryan Deschamps conducted preliminary research into what a digital history of 1990s Canada would look like. He found that despite evidence from print media and Usenet discussion boards that both the pro-independence and pro-separatist sides had important web presences, they were largely not preserved. Given the importance of the Quebec referendum to an understanding of modern Canadian history, this is a major loss. As Deschamps notes, "it was not clear to people in 1995 that web pages were historical documents that people might want to use for research in 2017 [or beyond]."[32] While the Bibliothèque et Archives nationales du Québec does have webpages from 1995, the sovereignty campaign pages were lost.[33]

It is distressing to think of the material that was lost from the early web that we do not know about. Early cultural sites? Homepages? Jokes? Academic servers? Wikipedia's "List of websites founded before 1995" is an overview of some of the kinds of early websites. As a crowd-sourced document, it is well positioned to draw on the collective memory of early web users.[34] Science museums, web comics, religious movements, campus newsletters, business websites, all feature on this list. The Exploratorium science museum in San Francisco, for example, opened its website in 1992, but the earliest snapshot we can access today dates from January 1997.[35] The *Economist*'s first website, launched in March 1994 by one of its correspondents, was reconfigured after eighteen months. As an *Economist* author bemoaned, "[a]ll records of the original website were subsequently lost. So much for the idea that the internet never forgets. It does."[36] Websites that lasted into 1996 or 1997 may have been preserved. Those that did not last that long, or were small enough that they eluded detection by web archives in their smaller early years, have not. If it was not for the people and institutions discussed in this book, we would have more stories like this. The gap would have been a lot longer. Ultimately, the Little Digital Dark Age of 1991–1996 illustrates how fortunate we are today.

Web archiving and digital preservation has come a long way between the early 1990s and 2001 and continues to develop today. The implications of this are still making themselves clear. As historians move into the web age of history, it will be important for them to know how the archives they use have been constructed and how they came into being. None of the web's memory is natural or intrinsic to the platform itself. It has been painstakingly constructed, the product of countless conscious decisions across Silicon Valley, national capitals, and academic institutions. Today, the Internet Archive is increasingly part of the internet's core infrastructure. As research libraries and national institutions preserve swaths of our cultural digital heritage online, it is critical to remember that none of this was inevitable. The web naturally forgets, and it is up to us to help it remember.

Introduction

1. Rothenberg, "Longevity of Digital Documents."
2. Hedstrom, "Digital Preservation."
3. Kuny, "Digital Dark Ages," 10.
4. Caruso, "In a Society Dependent on Technology."
5. McNeil, *Lurking*, 19.
6. Duffy and Flynn, "Iconic 9/11 News Coverage."
7. Ding, "Can Data Die?"
8. Gabriele and Perry, *Bright Ages*, 251.
9. Arnold, *What Is Medieval History?* 10
10. Starr, *Lost Enlightenment*.
11. Starr, *Lost Enlightenment*, 8.
12. Geary, *Phantoms of Remembrance*, 177.
13. Mayer-Schönberger, *Delete*, 52.
14. Goggin and McLelland, *Routledge Companion to Global Internet Histories*.
15. Schweig, *Renegade Rhymes*, 116.
16. See, for example, Chang et al., "Analysis of User Behaviour."
17. Balogun and Kalusopa, "Web Archiving of Indigenous Knowledge."
18. Lor and Britz, "A Moral Perspective."
19. See, for example, Cannon, "Bulgarian Collection" and Cannon, "Russian Collection."
20. Brügger, *Archived Web* and Milligan, *History in the Age of Abundance?*
21. For collections of case studies, see Brügger and Schroeder, *Web as History* and Brügger and Milligan, *SAGE Handbook of Web History*. Ethics have been well discussed in Lomborg, "Personal Internet Archives and Ethics"; Lomborg, "Ethical Considerations"; Lin et al., "We Could, but Should We?" Technical discussions are too many to cite, but see Ainsworth, Nelson, and Van de Sompel, "One Out of Five Archived Web Pages" and Nick Ruest et al., "Archives Unleashed Project." Finally, a hybrid of both perspectives is found in the comprehensive and thoughtful Gomes, Demidova, Winters, and Risse, *Past Web*.
22. Owens, *Theory and Craft*.
23. Two main exceptions are the work of Peter Webster and Jessica Ogden, both of

whom have engaged in interviews and are laying the foundation for a broader historical understanding of this field. See Webster, "Users, Technologies, Organisations" and Ogden, "Saving the Web." There is also an insider history in Kimpton and Ubois, "Year-by-Year."

24. Ovenden, *Burning the Books*, 214–215.

25. Rosenberg, "Search," 278–279.

26. In this, it is very much the opposite of what Michel Foucault lays out in his foundational *Archaeology of Knowledge*. Foucault holds that archeology "does not treat discourse as *document*, as a sign of something else, as an element that ought to be transparent, but whose unfortunate opacity must often be pierced . . . it is concerned with discourse in its own volume, as a *monument*." (155) Certainly, historians have had much to learn from postmodernist theory as they critically examine the link between sign and signal in the world around us. Fundamentally, however, I still believe that there is a good faith effort to be made to use these documents as reflections of what the world was.

27. Lori Emerson, *Reading Writing Interfaces*, xii. Similar narratives appear in the very helpful overview by Huhtamo and Parikka, "Introduction" and Ernst, "Media Archeography," 240.

28. Ernst, "Media Archeography," 240.

29. Chun, *Programmed Visions*, 80.

30. Kirschenbaum, *Bitstreams*.

31. Milligan, *Transformation of Historical Research*.

32. Kirschenbaum, *Bitstreams*, 29. Unless otherwise stated, all emphases are in the original.

33. Kirschenbaum, *Mechanisms*, 21.

34. Each interviewee was asked a series of semi-structured interview questions, and then had the opportunity to review the transcript and versions of this manuscript as I moved forward to publication.

Chapter 1 • Why the Web Could Be Saved

1. Yale, "History of Archives," 333.

2. Blouin and Rosenberg, *Processing the Past*, 9.

3. Walsham, "Social History of the Archive."

4. Milligan, *Transformation of Historical Research*.

5. Blouin and Rosenberg, *Processing the Past*, 10.

6. Saussy, "Appraising."

7. I am conscious that this seems a bit like I am constructing a mean-spirited strawman. Few historians are ignorant of the role that archives play in arguments, but this is rarely reflected—for a multitude of reasons, ranging from editorial pressures to our emphasis on well-crafted narrative—in published historical scholarship.

8. Cook, "Archive(s)," 504.

9. Cook, "Archive(s)," 511.

10. Friedrich, *Birth of the Archive*, 13 and 31.

11. As cited in Friedrich, *Birth of the Archive*, 167.

12. Friedrich, *Birth of the Archive*, 16.

13. For von Ranke, see Blouin and Rosenberg, *Processing the Past*, 16. Also see Peter Novick, *Noble Dream*.

14. Blouin and Rosenberg, *Processing the Past*, 38.

15. Fleckner, "F. Gerald Ham."

16. Ham, "Archival Edge," 13.

17. Interview with Edward Higgs, 3 February 2021. This anecdote is borne out in David Bearman, *Electronic Evidence* where historians do not appear at all.

18. Friedrich, *Birth of the Archive*, 6.

19. Of course, archival work is shaped by external forces: state mandates for government archives, board of director priorities for many community archives, and particular training received in graduate programs.

20. Blouin and Rosenberg, *Processing the Past*, 174–175.

21. Interview with Terry Kuny, 24 February 2021.

22. Ovenden, *Burning the Books*, 10.

23. Buckland, *Information and Society*, 6.

24. For a good overview of archival theory, see Ridener and Cook, *From Polders to Postmodernism*.

25. Buckland, *Information and Society*, 179.

26. This chapter is not the place to introduce the broad field, for which readers should turn to Owens, *Theory and Craft*.

27. Roy Rosenzweig, "What Is to Be Done?" 5 September 2003, Rosenzweig discussion at https://web.archive.org/web/20031104063602/http://www.historycooperative.org/phorum/read.php?f=14&i=2&t=2.

28. Lesk, "Preservation of New Technology," 2.

29. Lesk, "Preservation of New Technology," 14.

30. Task Force on Archiving of Digital Information, "Preserving Digital Information," iii and 3.

31. Pearson, "Importance of the Copy Census."

32. See https://www.lockss.org/about/history.

33. Discussed briefly in Vaidhyanathan, *Anarchist in the Library*, 15.

34. Cox, *First Generation of Electronic Records Archivists*, 50.

35. On the 1890 example, see National Historical Publications and Records Commission (NHPRC), "Research Issues," 1.

36. For example, see Hollier, "Archivist in the Electronic Age."

37. As claimed in Openness Advisory Panel, "Responsible Openness."

38. Adams and Brown, "Myths and Realities."

39. Lyons, "Preserving Electronic Government Information," 210.

40. Fishbein, "Appraising Information," 35.

41. Fishbein, "Appraising Information," 42.

42. NHPRC, "Research Issues in Electronic Records," 1.

43. Brown, "History of NARA's Custodial Program."

44. Higgs, "Historians, Archivists," 147.

45. For example, in Kirschenbaum, *Mechanisms*, 21.

46. Owens, *Theory and Craft*, 5.

47. NHPRC, "Research Issues," 1–2.

48. Kesner, "Automated Information Management," 163.
49. NHPRC, "Research Issues," 4.
50. Cox, *First Generation.*
51. Lesk, "Foreword," xvi.
52. Corrado and Moulaison, *Digital Preservation,* 21.
53. Lyons, "Preserving Electronic Government Information," 209.
54. Cox, *First Generation,* 24.
55. Cox, *First Generation,* 25.
56. Cox, *First Generation,* 194.
57. Hedstrom, *Archives & Manuscripts,* 9.
58. See an overview of this in Abbate, *Inventing the Internet.*
59. Driscoll, *Modem World,* 2.
60. Driscoll, *Modem World,* 35.
61. Driscoll, *Modem World,* 191.
62. Pew Research, "Online Use," 16 December 1996, https://www.pewresearch.org/politics/1996/12/16/online-use/.
63. Goggin and McLelland, "Internationalizing Internet Studies," 11.
64. Goggin and McLelland, "Introduction: Global Coordinates of Internet Histories," 5.
65. Goggin and McLelland, "Introduction: Global Coordinates of Internet Histories" 6.
66. Pargman and Palme, "ASCII Imperialism," 197–198.
67. Pargman and Palme, "ASCII Imperialism," 177.
68. Heffernan, *Magic and Loss,* 54–55.
69. Heffernan, *Magic and Loss.*
70. Rumsey, *When We Are No More,* 31.
71. Rayward, *Mundaneum,* 3. This book is a translation and adaptation of Raphaèle Cornille, Stéphanie Manfroid, and Manuela Valentino, *Le Mundaneum: les archives de la connaissance* (Brussels: Les impressions Nouvelles, 2008).
72. Rayward, *Mundaneum,* 13.
73. Rayward, *Mundaneum,* 15.
74. Wright, *Cataloguing the World,* 306.
75. Wells, "World Encyclopedia," as appears in Wells, *World Brain,* 13.
76. Buckland, "Emanuel Goldberg."
77. Bush, "As We May Think."
78. Wells, *World Brain,* 51.
79. Written in 1965, it was first published in 1967 as a chapter in Bush, *Science Is Not Enough.* It is also available as Bush, "Memex Revisited," in *New Media, Old Media.* That is the version cited and used in this chapter.
80. Bush, "Memex Revisited," 147.
81. Bush, "Memex Revisited," 157.
82. Chun, *Programmed Visions,* 80.
83. Nelson, "Complex Information Processing."
84. Barnet, *Memory Machines,* 76.
85. Barnet, *Memory Machines,* 85.
86. Barnet, *Memory Machines,* 86. See also the comprehensive overview at Nelson, "Xanalogical Structure."
87. Interview with Kahle, 26 February 2021.

88. Malloy, "Origins of Social Media."
89. Driscoll, *Modem World*, 25.
90. Driscoll, *Modem World*, 24.
91. Scott, "Textfiles.com."
92. Driscoll, *Modem World*, 100.
93. Scott, "Open Source, Open Hostility."
94. Rankin, *People's History of Computing*, 5–6.
95. Rankin, *People's History of Computing*, 246–247.
96. Dear, *Friendly Orange Glow*, xiii.
97. Milligan, *History in the Age of Abundance?* 71–72.
98. Friedrich, "Archivists," 313.

Chapter 2 • From Dark Age to Golden Age?

1. "Cracking Wise on the Internet," B8.
2. See "Decoding Common Internet Error Messages" for an example of this genre.
3. De Kosnik, *Rogue Archives*.
4. In a fascinating history, Erin Baucom traces the field as originating from challenges developing in the 1880s around punch cards and in the 1950s regarding magnetic tape, but at its core, the actual digital preservation movement arguably began, in her estimation, with the 1994 establishment of the Task Force on Archiving of Digital Information. See Baucom, "A Brief History," 5. Similarly, Hirtle's "History and Current State" notes early concerns around digital records in the 1960s at the University of Michigan and the UK National Archives, although he also traces the early stirrings of digital preservation per se around 1990 with work done between Xerox and Cornell at 1990.
5. Cook, "Easy to Byte, Harder to Chew," 204.
6. Cook, "Easy to Byte, Harder to Chew," 205.
7. Cook, "Easy to Byte, Harder to Chew," 206.
8. Waters, "Electronic Technologies."
9. Kenney and Personius, *Cornell/Xerox Commission on Preservation and Access*.
10. Task Force on Archiving of Digital Information, "Preserving Digital Information," iii.
11. Margaret Hedstrom, interview by author, 19 March 2021.
12. Task Force on Archiving of Digital Information, "Preserving Digital Information," 1.
13. Task Force on Archiving of Digital Information, "Preserving Digital Information," 1–2.
14. Task Force on Archiving of Digital Information, "Preserving Digital Information," 18.
15. Hedstrom, "Digital Preservation," 191–92.
16. Hedstrom, interview by author, 19 March 2021.
17. Kuny, "A Digital Dark Ages?"
18. Kuny, interview by author, 24 February 2021.
19. Gitelman, *Always Already New*, 1.
20. Brabazon, "Dead Media."
21. Sterling, "The Life and Death of Media."

22. Sterling, "The Life and Death of Media."
23. Sterling, "The Life and Death of Media."
24. Sterling, "The Life and Death of Media."
25. Sterling, "The Life and Death of Media."
26. Sterling and Kadrey, "Dead Media Project."
27. Brabazon, "Dead Media."
28. Bak, "Dead Media Project."
29. Brabazon, "Dead Media."
30. Bak, "Dead Media Project."
31. Rothenberg, "Longevity of Digital Documents."
32. Kuny, "A Digital Dark Ages?"
33. Caruso, "Society Dependent on Technology."
34. Terry Kuny, interview by author, 24 February 2021.
35. Higgs, interview by author, 3 February 2021.
36. Hedstrom, interview by author, 19 March 2021.
37. Koerbin, interview by author, 1 March 2021.
38. Koerbin, in email discussion with the author, 11 March 2021.
39. Lessig, *Free Culture*, 100.
40. Brand, *Clock of the Long Now*, 84.
41. Brand, *Clock of the Long Now*, 79.
42. Hedstrom, interview by author, 19 March 2021.
43. Kahle, interview by author, 26 February 2021.
44. Hedstrom, interview by author, 19 March 2021.
45. Hedstrom, interview by author, 19 March 2021.
46. Caruso, "Can the Web be a Time Capsule."
47. Auletta, "The Microsoft Provocateur."
48. Auletta, "The Microsoft Provocateur."
49. Myhrvold, "Road Kill."
50. Stross, *Microsoft Way*; Clark, *Netscape Time*.
51. Myhrvold, "Road Kill."
52. Myhrvold, "Road Kill."
53. Stross, *Microsoft Way*, 136 and 216.
54. Caruso, "Can the Web Be a Time Capsule."
55. Caruso, "Can the Web Be a Time Capsule."
56. Milligan, *History in the Age of Abundance?*
57. Berreby, "The Art of Losing."
58. Markoff, "When Big Brother Is a Librarian."
59. Barnes, "Nothing but Net."
60. Gardner, "Report on Documenting the Digital Dark Age." My thanks to Margaret Hedstrom for providing me with this document.
61. Gardner, "Report on Documenting the Digital Dark Age," 3.
62. Gardner, "Report on Documenting the Digital Dark Age," 3.
63. My thanks to Margaret Hedstrom for making her records from the conference available to me, cited as Margaret Hedstrom Personal Collection. I have uploaded these documents with her permission to https://archive.org/details/documenting-the-digital-age/.

64. Hedstrom, interview by author, 19 March 2021.
65. Waters, "Choices in Digital Archiving."
66. Barnes, "Nothing but Net."
67. Rosenzweig, "Scarcity or Abundance?" 740 and 758.
68. Documenting the Digital Dark Age, "Documenting the Digital Age."
69. Hedstrom, interview by author, 19 March 2021.
70. Kahle, "Archiving the Internet," *Scientific American*.
71. Hedstrom, "How Do We Make Electronic Archives."
72. Waters, "How Do We Archive Digital Records?"
73. Myhrvold, "Why Archive the Internet?"
74. Hedstrom, interview by author, 19 March 2021.
75. Myhrvold, "Internet History Conference"; included as appendix to Gardner, "Report on Documenting the Digital Dark Age," 5.
76. Gardner, "Report on Documenting the Digital Dark Age," 9–11.
77. Myhrvold, "Internet History Conference," 3.
78. As a historian, of course, my opinion is that there can never be too many historians.
79. Barnes, "Nothing but Net."
80. Hedstrom Personal Collection, Letter from Philip L. Cantelon to Margaret Hedstrom, 21 February 1997 and Fax from James B. Gardner to Margaret Hedstrom, 5 May 1997.
81. "Film on Preservation"; airing date included in Wallich, "Preserving the Word."
82. Sanders, dir., *Into the Future*.
83. Sanders, dir., *Into the Future*.
84. Sanders, dir., *Into the Future*.
85. Wallich, "Preserving the Word."
86. Dick, "Into the Future."
87. Skillman, "Video Review," 33.
88. Sudhir, "Preserving the Past."
89. Budd, "Review."
90. Ogden, "Review."
91. Davis and MacLean, "Mapping the Project," 5.
92. For example, see Lovejoy, "Artists' Books" and McLaughlin, "The Art Site."
93. Mayfield, "How to Preserve Digital Art."
94. "Participants."
95. Hillis, "Millennium Clock."
96. Hillis, "Millennium Clock."
97. "The Rosetta Project," Long Now Foundation, https://rosettaproject.org/.
98. "The Long Server Project," Long Now Foundation, http://longserver.org.
99. Nowviskie, "Digital Humanities," i7.
100. Brand, *Clock of the Long Now*, 87.
101. Davis and MacLean, "Mapping the Project."
102. Lyman and Besser, "Defining the Problem," 11.
103. MacLean, "Setting the Stage," 32–33.
104. Brand, *Clock of the Long Now*, 103.
105. Nowviskie, "Digital Humanities."

106. MacLean, "Setting the Stage," 33.
107. Brand, *Clock of the Long Now*, 91.
108. MacLean, "Setting the Stage," 66.
109. MacLean, "Setting the Stage," 34.
110. MacLean, "Setting the Stage," 34.
111. "Public Session," 38.
112. "Public Session," 39–40.
113. "Public Session," 42.
114. Crymble, *Technology and the Historian.*
115. Swierenga, "Clio and Computers."
116. Gaffield, "Clio and Computers in Canada."
117. Fishbein, "Welcome," 4.
118. Rhoads, "Preface," xi.
119. Lemercier and Zalc, *Quantitative Methods*, 14–17.
120. Higgs, interview by author, 3 February 2021.
121. Thanks to Peter Webster for this thought.
122. Gaffield, "Clio and Computers in Canada."
123. Higgs, interview by author, 3 February 2021.
124. "Editorial," *History and Computing*, ii.
125. "Editorial," *History and Computing*, ii.
126. "Editorial," *History and Computing*, iii.
127. "Editorial," *History and Computing*, ii.
128. Zweig, "Virtual Records," 175.
129. Zweig, "Virtual Records," 176–177.
130. Zweig, "Virtual Records," 180.
131. Higgs, interview by author, 3 February 2021.
132. Hedstrom, interview by author, 19 March 2021.
133. Ross, "Historians," 1.
134. Ross, "Historians," 6.
135. Schürer, "Outlook of the User Community," 246.
136. Schürer, "Outlook of the User Community," 247.
137. Schürer, "Information Technology," 294.
138. Morris, "Electronic Documents," 302.
139. Morris, "Electronic Documents," 304.
140. Morris, "Electronic Documents," 311.
141. Ross, "Introduction," xxi.
142. Kenny, "Keynote," 8–9.
143. Hedstrom, "Electronic Archives," 91.
144. Ross, "Historians," 6.
145. Higgs, interview by author, 3 February 2021.
146. As of 2023, the piece had been cited 367 times according to Google Scholar. This is a very high number for historical scholarship (especially as digital history is still a relatively small scholarly subfield).
147. "About," Roy Rosenzweig Center for History and New Media Website, 2021, https://web.archive.org/web/20211202230039/https://rrchnm.org/about/.
148. For more, see https://rrchnm.org/.

149. Rosenzweig, "Scarcity or Abundance?" 738.
150. Rosenzweig, "Scarcity or Abundance?" 745–746.
151. Rosenzweig, "Scarcity or Abundance?" 752.
152. Rosenzweig, "Scarcity or Abundance?" 756.
153. Rosenzweig, "Scarcity or Abundance?" 762.
154. "AHR Forum Essay," 734.
155. The discussion was hosted by HistoryCooperative.org, which had been a digital intermediary between scholarly associations that published journals and readers (often with a print embargo, before simultaneous print/digital). For more on this, see Katz's "Publishing History."
156. Antoinette Burton, "RE: archivists v. historians? medieval v. modern?" 15 September 2003, Rosenzweig discussion at https://web.archive.org/web/20031104063602/http://www.historycooperative.org/phorum/read.php?f=14&i=2&t=2.
157. Roy Rosenzweig, "Preservation through Neglect," 11 September 2003, Rosenzweig discussion at https://web.archive.org/web/20031104063602/http://www.historycooperative.org/phorum/read.php?f=14&i=2&t=2.

Chapter 3 • Building the Universal Library

1. Steinberg, "Seek and Ye Shall Find."
2. Kahle, "Universal Access."
3. "Lease Space in the Presidio: 116 Sheridan Avenue," as linked from https://web.archive.org/web/20200315210742/https://www.presidio.gov/presidio-workspaces-leasing. Unfortunately, the live page disappeared between May 2020 and July 2021, and the author neglected to download a snapshot. How embarrassing for a book about web preservation.
4. Kahle, interview by author, 26 February 2021.
5. Rosenberg, "Search," 273.
6. O'Brien, "He Fights," BU1. Also discussed in Livingston, *Founders at Work*, 265.
7. Kahle, interview by author, 26 February 2021.
8. Garsson, "Artificial Intelligence," 48.
9. Alexander, "Parallel Computer," 20.
10. "Thinking Machines in Chapter 11." See also Markoff, "Supercomputer Pioneer."
11. Flynn, "A New Power."
12. Markoff, "For Shakespeare."
13. Markoff, "For Shakespeare."
14. As quoted in Ogden, "Saving the Web," 79.
15. Livingston, *Founders at Work*, 267.
16. Kahle, interview by author, 26 February 2021.
17. Markoff, "For Shakespeare" and Lukanuski, "Help Is on the WAIS," 743.
18. Lukanuski, "Help Is on the WAIS," 742.
19. Markoff, "For Shakespeare."
20. Kahle, interview by author, 26 February 2021.
21. Livingston, *Founders at Work*, 271.
22. Livingston, *Founders at Work*, 269.
23. Turner, *Counterculture to Cyberculture*, 208–209.
24. Livingston, *Founders at Work*, 269.

25. "WAIS, Inc. Plan 95."
26. Churbuck, "Good-bye, Dewey decimals."
27. "WAIS Inc. Releases New Network."
28. The WAIS Collection at the Internet Archive provides comprehensive documentation on WAIS Inc (https://archive.org/details/wais). For example, see Letter from Brewster Kahle to WAIS Employees, 18 June 1994, available at https://archive.org/details/wais.
29. Letter from James H. Billington to Brewster Kahle, 20 July 1994, available at https://archive.org/details/04Kahle000826.
30. Kahle, interview by author, 26 February 2021.
31. Fax from Brewster Kahle to David Cole, 11 January 1995, available at https://archive.org/details/09Kahle000639.
32. "America Online to Buy Two Firms," *United Press International*, 22 May 1995.
33. Kahle, interview by author, 26 February 2021.
34. Email from Brewster Kahle to WAIS staff, 24 August 1995, available at https://archive.org/details/09Kahle002620; Email from D. Kaiser to Brewster Kahle, 15 September 1995, available at https://archive.org/details/05Kahle001064.
35. Email from Brewster Kahle to Steve Case, 25 August 1995, available at https://archive.org/details/09Kahle002614
36. Email from David Lytel to Ted Leonsis, 5 December 1995, available at https://archive.org/details/09Kahle002606
37. Email from Brewster Kahle to WAIS Staff, 26 January 1996, available at https://archive.org/details/09Kahle002526 and email from Brewster Kahle to Steve Case, 18 February 1996, available at https://archive.org/details/09Kahle002526.
38. Email from Brewster Kahle to Mike Connors, 8 November 1995, available at https://archive.org/details/04Kahle000006.
39. Email from Brewster Kahle to Deanna, 17 November 1995, available at https://archive.org/details/04Kahle000003.
40. Robin, "WAIS—A New Vision." Available at https://archive.org/details/05Kahle001743.
41. Cunningham, "Who Archives the World Wide Web?" 18.
42. Email between Brewster Kahle and Carl Malamud, 9 and 10 January 1996, available at https://archive.org/details/CM.BK.MailTapeDrive.
43. There is some disagreement as to whether it was founded in March 1996 (as per Berreby, "The Art of Losing") or April 1996 as per Kimpton and Ubois, "Year-by-Year," 202. Kahle noted April 1996 in our interview. As we see in this chapter, the Internet Archive's activities had in any event begun before its formal incorporation.
44. Foot and Schneider, *Web Campaigning*, 8.
45. Foot and Schneider, *Web Campaigning*, 9.
46. Kahle, interview by author, 26 February 2021.
47. Letter from Spencer R. Crew to Alan Keyes (Presidential Candidate), 28 February 1996, available at https://archive.org/details/04Kahle000461.
48. Fax from Leila Murphy (National Museum of American History) to Brewster Kahle, 4 March 1996, available at https://archive.org/details/04Kahle000476.
49. Smithsonian Institute, "National Museum of American History."
50. Berreby, "The Art of Losing."

51. Kahle, interview by author, 26 February 2021.
52. Kimpton and Ubois, "Year-by-Year," 203.
53. A photocopy of the check written by AOL to him for $2,253,400 in October 1995 was in his archive, and the media generally reported that he had made around $15 million from the sale. See, for example, "Proceeds from Sale of AOL Stock Cheque."
54. Kahle, "Archiving the Internet: Towards a Core Internet Service."
55. Selingo, "Attempting to Archive the Entire Internet."
56. Kahle, "Archiving the Internet."
57. Kahle, "Archiving the Internet."
58. Steinberg, "Seek and Ye Shall Find."
59. Dunleavey, "Going Giddy."
60. Kahle, "The Internet Companion Company."
61. Kahle, "The Internet Companion Company."
62. Deger, "Internet Historian" and Flynn, "Alexa Internet," as found in "Alexa Internet Press Clips," https://archive.org/details/alexainternetpre19unse_1/page/n39/mode/2up.
63. Selingo, "Attempting to Archive the Entire Internet."
64. Kahle, interview by author, 26 February 2021. By 2020, the Internet Archive crawling was so mature that the dataflow from Alexa was no longer necessary.
65. Kahle, interview by author, 26 February 2021.
66. "Alexa Internet Introduces Web Navigation."
67. Ubois, "It's a Jungle," 150.
68. McKenzie, "Alexa Gets You."
69. Ankerson, "Take Me Back!"
70. "Alexa Internet Introduces Web Navigation."
71. Cramb, "Cyber Historian."
72. Markoff, "New Service."
73. Livingston, *Founders at Work*, 274–275.
74. Weise, "Tracking the Web."
75. Chandrasekaran, "In California, Creating a Web of the Past."
76. Kahle, interview by author, 26 February 2021.
77. Foot, interview by author, 23 March 2021.
78. Moozakis, "Daunting Task," 31.
79. Cunningham, "Brewster's Millions," 18.
80. Kahle, interview by author, 26 February 2021.
81. Schneider, interview by author, 23 March 2021.
82. Kahle, email with author, 29 February 2024.
83. Burner and Kahle, "Arc File Format."
84. Christian, "Why Does Amazon."
85. Williams, "Alexa Makes Sense."
86. Livingston, *Founders at Work*, 276.
87. Helm, "Amazon to Buy 3 More Internet Firms."
88. "Amazon.com to Pay $450-Million."
89. O'Brien, "Online Librarian," 55.
90. Markoff, "Data-Gathering Program," Computer 5.
91. Gray, "Amazon Unit Faces $1.9 Million Payout."

92. Cunningham, "Brewster's Millions."
93. Marcus, "US Starts Archive."
94. Cunningham, "Who Archives the World Wide Web?" 18.
95. Berreby, "The Art of Losing."
96. Gillmor, "Internet's Archive Will Preserve the Web," 12.
97. Chandrasekaran, "In California, Creating a Web of the Past."
98. Markoff, "When Big Brother Is a Librarian."
99. Steinberg, "Seek and Ye Shall Find."
100. Steinberg, "Seek and Ye Shall Find."
101. Lane, *Naked in Cyberspace*. See coverage in venues such as Quittner, "Invasion Of Privacy."
102. Miller, "You Are a Database."
103. Chandrasekaran, "In California, Creating a Web of the Past."
104. Miller, "You Are a Database."
105. Kimpton and Ubois, "Year-by-Year," 203.
106. "The 7-Terabyte Man."
107. Jackson, "Archive Holds Wealth of Data." "Several minutes" was noted in Manning, "Alexa Brings Dead Webs Back to Life."
108. Kimpton and Ubois, "Year-by-Year," 205 and National Public Radio, "Problems of Digital Preservation."
109. "Proposing a Project or a Donation."
110. Markoff, "When Big Brother Is a Librarian."
111. Markoff, "When Big Brother Is a Librarian."
112. Kimpton and Ubois, "Year-by-Year," 207; Schwartz, "A Library of Web Pages."
113. Kahle, interview by author, 26 February 2021.
114. Schwartz, "A Library of Web Pages."
115. Kornblum, "Web-Page Database."
116. Mieszkowski, "Dumpster Diving."
117. Notess, "The Wayback Machine."
118. Goel, "Beta Wayback Machine."
119. Hickey, "The Wayback Machine."
120. Aufderheide and Jaszi, *Reclaiming Fair Use*, 6.
121. Aufderheide and Jaszi, *Reclaiming Fair Use*, 18.
122. Kahle, interview by author, 26 February 2021.
123. Johns, *Piracy*, 479.
124. Johns, *Piracy*, 463.
125. Johns, *Piracy*, 483.
126. Kahle, interview by author, 26 February 2021.
127. Schwartz, "A Library of Web Pages."
128. Mieszkowski, "Dumpster Diving."
129. Koster, "ANNOUNCE." See also Milligan, *History in the Age of Abundance?* 92–93.
130. Kimpton and Ubois, "Year-by-Year," 203.
131. Quote from an Internet Archive spokesperson quoted in Bixenspan, "When the Internet Archive Forgets." See also Kimpton and Ubois, "Year-by-Year," 203.
132. "The Oakland Archive Policy."

133. The Oakland Archival Policy was "developed with help and advice of representatives of the Electronic Frontier Foundation, Chilling Effects, The Council on Library and Information Resources, the Berkeley Boalt School of Law, and various other commercial and noncommercial organizations through a meeting held by the Archive Policy Special Interest Group (SIG), an ad hoc, informal group of persons interested the practice of digital archiving."

134. "The Oakland Archival Policy."

135. "The Oakland Archival Policy."

136. Lessig, *Free Culture,* 114.

137. Lessig, *Free Culture,* 213.

138. "Copyright Term Extension."

139. Lessig, *Free Culture,* 215.

140. "The Constitution of the United States."

141. Lessig, *Free Culture,* 228.

142. Lessig, *Free Culture,* 228.

143. Koman, "Riding Along."

144. Kahle, interview by author, 26 February 2021.

145. Koman, "Riding Along."

146. Koman, "Riding Along."

147. Koman, "Riding Along."

148. Koman, "Riding Along."

149. Koman, "Riding Along."

150. Solomon, "He Crosses Country," 58.

151. Koman, "Riding Along."

152. Lessig, *Free Culture,* 239.

153. Carlson and Young, "Google Will Digitize," 37.

154. Johns, *Piracy,* 511.

155. Boutin, "Archivist."

156. Founded by Kahle, the OCA included the Internet Archive and Yahoo! Books would be provided by the University of California system, the University of Toronto, the European Archive, and the UK National Archives, with additional technical support from HP Labs, Adobe, and tech publisher O'Reilly. See Berr, "Bookworms."

157. Berr, "Bookworms."

158. Hafner, "Challenge to Google."

159. Hafner, "Microsoft to Offer Online Book-Content."

160. Young, "Scribes of the Digital Era," 34.

161. Kelly, "Scan This Book!"

162. Borsuk, *The Book,* 218.

163. Streitfeld, "In a Flood Tide of Digital Data."

164. Borsuk, *The Book,* 219.

165. Rosenzweig, "Scarcity or Abundance?" 752.

166. Rosenzweig, "Scarcity or Abundance?" 752.

167. Ovenden, *Burning the Books,* 203.

168. All tax filings drawn from ProPublica's Nonprofit Explorer for the Internet Archive, available at https://projects.propublica.org/nonprofits/organizations/9432 42767.

169. Farrell et al., "Results of a Survey."
170. Srinivasan, "The Internet Archive."
171. Popper, "Dream of New Kind."
172. B., "Internet Archive Addresses Housing Crisis."

Chapter 4 • From Selective to Comprehensive

1. National Research Council, *LC21*, 2–4.
2. Delsey, "National Library's Role," 89.
3. Kuny, interview by author, 24 February 2021.
4. "Preserving the WorldWideWeb." Discussed in email from Paul Koerbin, 11 March 2021.
5. Hedstrom, interview by author, 19 March 2021.
6. Gomes, Demidova, Winters, and Risse, "Prologue," *The Past Web*, xi.
7. Masanès, "Selection," 76.
8. Never ask a historian what to collect (me included). They often say "everything."
9. Delsey, "National Library's Role," 89.
10. Feijen, "DNEP."
11. See the overview in Lunn, "National Library of Canada."
12. Cleveland, "Overview of Document Management."
13. Hedstrom, interview by author, 19 March 2021.
14. Nilsen, "Government Information Policy," 191.
15. Nilsen, "Government Information Policy," 205.
16. Hodges and Lunau, "National Library of Canada."
17. Delsey, "National Library's Role," 89.
18. Delsey, "National Library's Role," 89.
19. Delsey, "National Library's Role," 89.
20. Lilleniit, "National Library of Canada," 28.
21. Electronic Publications Pilot Project (EPPP), *Electronic Publications Pilot Project*, 8.
22. Kuny, interview by author, 24 February 2021.
23. Brodie, "National Library's Electronic Publications," 1.
24. Kuny, interview by author, 24 February 2021.
25. EPPP, *Electronic Publications Pilot Project*, 11.
26. Kuny, interview by author, 24 February 2021.
27. EPPP, *Electronic Publications Pilot Project*, 9.
28. EPPP, *Electronic Publications Pilot Project*, 20.
29. Lilleniit, "National Library of Canada," 33
30. EPPP, *Electronic Publications Pilot Project*, 20.
31. EPPP, *Electronic Publications Pilot Project*, 5.
32. Kuny, interview by author, 24 February 2021.
33. Government of Canada, *Library and Archives of Canada Act*.
34. Milligan and Smyth, "Studying the Web," 47.
35. Milligan and Smyth, "Studying the Web," 52.
36. Mannerheim, "Preserving the Digital Heritage," 2.
37. Mannerheim, "Preserving the Digital Heritage," 4.
38. Arvidson and Persson, "Kulturarw3."
39. This was the first of a series of comprehensive "reference group" meetings, with

minutes documented and kept in the Internet Archive. As I cannot read Swedish, I used Google Translate to translate these meeting minutes.

40. "Idea Hatchery Colloquium."
41. "Reference Group Meeting," 17 September 1996.
42. "Statistics."
43. See also Herres, "Project Kulturarw3," 33.
44. Kahle, interview by author, 26 February 2021.
45. "Reference Group Meeting," 14 May 1997.
46. "Reference Group Meeting," 14 May 1997.
47. "Reference Group Meeting," 10 September 1997.
48. Arvidson and Persson, "Kulturarw3."
49. Government of Sweden, *Tryckfrihetsförordningen*.
50. Swedish Authority for Privacy Protection, "News in English."
51. "Reference Group Meeting," 11 February 1997.
52. "Reference Group Meeting," 14 May 1997.
53. Mannerheim, "Preserving the Digital Heritage," 3.
54. Mannerheim, "Preserving the Digital Heritage," 3.
55. "Reference group meeting," 17 September 1998.
56. "Reference group meeting," 8 May 2000.
57. Herres, "Project Kulturarw3," 32.
58. Lidman, "Ny förordning för Kulturarvsprojektet."
59. Herres, "Project Kulturarw3," 71.
60. "Reference Group Meeting," 10 September 1997.
61. "Reference Group Meeting," 3 April 1998.
62. "Reference Group Meeting," 25 November 1999.
63. Nordic Web Archive, "About the NWA Toolset."
64. Herres, "Project Kulturarw3," 72.
65. Smith, "Still Lost in Cyberspace?" 279.
66. For rich context, see Hegarty, "Invention of the Archived Web."
67. Butler, "Libraries Break Down Walls," 55.
68. Phillips and Koerbin, "PANDORA," 30.
69. Hegarty, "Invention of the Archived Web," 7–8.
70. Koerbin, email, 11 March 2021.
71. Yiacoum, "In a Pickle," 4.
72. Sinclair, "Web Wise," 6.
73. Broekhuyse, "Rescuers Save Endangered Websites," 43.
74. Koerbin, interview by author, 1 March 2021.
75. Sinclair, "Web Wise," 6.
76. Koerbin, interview by author, 1 March 2021.
77. Koerbin, email, 11 March 2021.
78. Hegarty, "Invention of the Archived Web," 12.
79. Hegarty, "Invention of the Archived Web," 10–11.
80. Yiacoum, "In a Pickle," 4.
81. National Library of Australia, "Selection Committee."
82. Koerbin, interview by author, 1 March 2021.
83. National Library of Australia, "Selection Committee."

84. Koerbin, interview by author, 1 March 2021.
85. Koerbin, interview by author, 1 March 2021.
86. Hegarty, "Invention of the Archived Web," 11.
87. National Library of Australia, "Selection Committee" and Koerbin, interview by author, 1 March 2021.
88. Yiacoum, "In a Pickle," 4.
89. Koerbin, interview by author, 1 March 2021.
90. National Library of Australia, "National Strategy for Provision of Access."
91. Koerbin, interview by author, 1 March 2021.
92. Koerbin, interview by author, 1 March 2021.
93. See discussion of lobbying in Hegarty, "Invention of the Archived Web," 6.
94. Koerbin, email, 11 March 2021.
95. Koerbin, interview by author, 1 March 2021.
96. Koerbin, interview by author, 1 March 2021.
97. See "radical incrementalism" being used by the National Library of Australia's director-general in 2019 in Easton, "Australia's Top Librarian." This is considered part of a broader discussion of the evolution of the web archive in Koerbin, "National Web Archiving," 30.
98. Broekhuyse, "Rescuers Save Endangered Websites," 43.
99. National Library of Australia, "National Strategy for Provision of Access."
100. National Library of Australia, "Guidelines for the Selection."
101. National Library of Australia, "Guidelines for the Selection."
102. National Library of Australia, "Guidelines for the Selection."
103. Hegarty, "Invention of the Archived Web," 11.
104. Phillips, "The National Library of Australia."
105. Martin, "Where Websites Go to Die," 14.
106. Morrison, "Www.Nla.Gov.Au/Pandora," 274.
107. Koerbin, interview by author, 1 March 2021.
108. Pasieczny, "A Goldmine in the Sky," 11.
109. Koerbin, email, 11 March 2021.
110. Phillips and Koerbin, "PANDORA," 20.
111. Burgess, "Club Buggery Joins Wiggles," A8.
112. "FED: Foreign Affairs Website."
113. Broekhuyse, "Rescuers Save Endangered Websites," 43.
114. Yiacoum, "In a Pickle," 4.
115. Meacham, "Open All Hours," 6.
116. Dickins, "Library's Porn Plan," 23.
117. Martin, "Where Websites Go to Die," 14.
118. Morrison, "Www.Nla.Gov.Au/Pandora," 275.
119. Morrison, "Www.Nla.Gov.Au/Pandora," 275.
120. National Library of Australia, "What We Collect: Ephemera."
121. Koerbin, interview by author, 1 March 2021.
122. Martin, "Where Websites Go to Die," 14.
123. Morrison, "Www.Nla.Gov.Au/Pandora," 276.
124. Koerbin, interview by author, 1 March 2021.
125. Koerbin, email, 11 March 2021.

126. Grose, "Staff Harvest Web for History."
127. "Copyright Act 1968—Sect 200AB."
128. Koerbin, in email discussion with the author, 4 July 2023.
129. Smith, "Still Lost in Cyberspace?" 279.
130. Bruns, "Contemporary Culture and the Web."
131. Grose, "Staff Harvest Web for History."
132. Koerbin, email, 4 July 2023.
133. Smith, "Still Lost in Cyberspace?" 280.
134. Koerbin, "National Web Archiving," 27.
135. Meacham, "Editing Our Future," 18.
136. Meacham, "Editing Our Future," 18.
137. Meacham, "Editing Our Future," 18.
138. Koerbin, "National Web Archiving," 29.
139. Grose, "National Library Stocktaking."
140. "We're Big on Virtual Presence."
141. Koerbin, interview by author, 1 March 2021.
142. Stephens, "Caught in The Web,"A2.
143. Koerbin, email, 4 July 2023.
144. "Archived Websites," *Trove*, accessed 7 June 2021, https://trove.nla.gov.au/website?q=.
145. "Alexa Internet Donates Archive."
146. "Alexa Internet Donates Archive."
147. "Alexa Internet Donates Archive."
148. Grotke, interview by author, 5 March 2021.
149. The Library of Congress is a de facto national library, despite formally being the library of the American legislative branch. Despite this technical ambiguity, it clearly fills the role of a national library. The Library of Congress provides a national catalogue, exercises a power akin to legal deposit, and is a member of international bodies alongside other national libraries. See National Research Council, *LC21*, 152.
150. Hafner, "Saving the Nation's Digital Legacy," G1.
151. Hedstrom, interview by author, 19 March 2021.
152. National Research Council, *LC21*, 2–4.
153. National Research Council, *LC21*, 100.
154. National Research Council, *LC21*, 87.
155. National Research Council, *LC21*, 89.
156. Hafner, "Saving the Nation's Digital Legacy," G1.
157. Martin, "Where Websites Go to Die," 14.
158. Hedstrom, interview by author, 19 March 2021.
159. Hedstrom, interview by author, 19 March 2021.
160. Hedstrom, interview by author, 19 March 2021.
161. Grotke, interview by author, 5 March 2021.
162. Arms, "Web Preservation Project: Interim Report."
163. Arms, "Web Preservation Project: Interim Report."
164. Arms, "Web Preservation Project: Interim Report."
165. Grotke, interview by author, 5 March 2021.
166. Arms, "Web Preservation Project: Final Report."

167. Kahle, interview by author, 26 February 2021.
168. Grotke, interview by author, 5 March 2021.
169. Grotke, interview by author, 5 March 2021.
170. Arms, "Web Preservation Project: Final Report," 7.
171. "Internet Library Enables Users." Apparently, according to the actual web archived version of the collection, one could navigate a subset of 797 sites and 800GB of data via the portal itself. See "Election 2000 Collection."
172. "Election 2000, as It Happened."
173. Meddis, "Lots of Good Things."
174. Grotke, interview by author, 5 March 2021.
175. Schneider, interview by author, 23 March 2021.
176. Grotke, interview by author, 5 March 2021.
177. As noted in Cannon, "The Bulgarian Collection."
178. See the list of collections at https://www.loc.gov/web-archives/collections/?sp=1.
179. Kurutuz, "Meet Your Meme Lords."
180. That said, as the Long Now Foundation encourages us to think, national libraries are certainly not assured to survive over the next thousands of years.
181. Milligan, *History in the Age of Abundance?* 52.
182. Milligan, *History in the Age of Abundance?* 52–53.
183. Kimpton and Ubois, "Year-by-Year," 210.
184. International Internet Preservation Consortium, "About the Consortium."

Chapter 5 • Archiving Disaster

1. "Summary Video of Key Events," Understanding 9/11.
2. Rainie and Kalsnes, "Commons of the Tragedy," 14.
3. Rainie and Kalsnes, "Commons of the Tragedy," 10.
4. National Research Council, *Internet Under Crisis Conditions*, 29.
5. National Research Council, *Internet Under Crisis Conditions*, 41.
6. National Research Council, *Internet Under Crisis Conditions*, 42.
7. Reprinted at Wendland, "Overloaded Internet."
8. Allan, "Reweaving the Internet," 123.
9. Allan, "Reweaving the Internet," 125.
10. Lessig, *Free Culture*, 40.
11. National Research Council, *Internet Under Crisis Conditions*, 37–38.
12. National Research Council, *Internet Under Crisis Conditions*, 47.
13. Bruno, "Archiving Anguish."
14. National Research Council, *Internet Under Crisis Conditions*, 10.
15. "Special Release: Post–September 11."
16. "Special Release: Post–September 11."
17. Harmon, "The Talk Online."
18. Garreau, "A Shaken Global Village."
19. "safe.millennium.berkeley.edu," archived 14 September 2001, available via the Internet Archive Wayback Machine at https://web.archive.org/web/20010914180341/http://safe.millennium.berkeley.edu/.

20. The best comprehensive report on Internet responses can be found at Rainie and Kalsnes, "Commons of the Tragedy."
21. "Special Release: Post–September 11."
22. Pew Internet and American Life Project, "One Year Later," 5.
23. Pew Internet and American Life Project, "One Year Later," 12.
24. "America under Attack."
25. Cox et al., "The Day the World Changed."
26. Rivard, "Collecting Disaster," 93.
27. Grotke, "Ask the Recommending Officer."
28. Grotke, interview by author, 5 March 2021.
29. Grotke, interview by author, 5 March 2021.
30. Foot, interview by author, 23 March 2021.
31. Schneider, interview by author, 23 March 2021.
32. Foot, interview by author, 23 March 2021.
33. Grotke, interview by author, 5 March 2021.
34. Grotke, "Ask the Recommending Officer."
35. Grotke, "Ask the Recommending Officer."
36. Grotke, interview by author, 5 March 2021.
37. Kresh, "Courting Disaster," 155.
38. Kresh, "Courting Disaster," 155.
39. "ADL Web Site Included."
40. Foot, interview by author, 23 March 2021.
41. Grotke, interview by author, 5 March 2021.
42. Schneider, interview by author, 23 March 2021.
43. Foot, interview by author, 23 March 2021.
44. "Please Help Us Build a Web Archive."
45. "Please Help Us Build a Web Archive."
46. Foster, "2 Scholars Archive Web Sites."
47. Mirapaul, "How the Net Is Documenting a Watershed Moment."
48. Schneider, interview by author, 23 March 2021.
49. Mirapaul, "How the Net Is Documenting a Watershed Moment."
50. "September 11 Web Archive."
51. Grotke, interview by author, 5 March 2021.
52. Grotke, interview by author, 5 March 2021.
53. Foot, interview by author, 23 March 2021.
54. Maemura, Worby, Milligan, and Becker, "If These Crawls Could Talk."
55. Scheinfeldt, interview by author, 8 April 2021.
56. Rosenzweig, Sparrow, and Cohen, "About Us."
57. Gardner and Henry, "September 11 and the Mourning After," 38–39.
58. "911history.net."
59. Brier, "The Intentional Archive."
60. Scheinfeldt, interview by author, 8 April 2021.
61. "A. P. Sloan Foundation Grants $700,000."
62. Salmon, "Digital 9/11 Project."
63. Pymm, "Archives and Web 2.0," 19.

64. "Saving the Histories of September 11, 2001."
65. Suzukamo, "E-Mail Archive."
66. "Stories of September 11."
67. Rivard, "Archiving Disaster." *Academia.edu.*
68. "An Introduction," *Where Were You . . .*, accessed 6 July 2021, http://wherewereyou.org.
69. Scheinfeldt, interview by author, 8 April 2021.
70. Rivard, "Archiving Disaster."
71. Scheinfeldt, interview by author, 8 April 2021.
72. Scheinfeldt, interview by author, 8 April 2021.
73. Scheinfeldt, interview by author, 8 April 2021.
74. Parman, "Electronic Archive."
75. On the media impact, Brier and Brown, "The September 11 Digital Archive," 105. Data on Google from Pymm, "Archives and Web 2.0," 20.
76. Scheinfeldt, interview by author, 8 April 2021.
77. "September 11: Bearing Witness to History."
78. Vallis, "Ordinary Items."
79. Scheinfeldt, interview by author, 8 April 2021.
80. Scheinfeldt, interview by author, 8 April 2021.
81. Scheinfeldt, interview by author, 8 April 2021.
82. Scheinfeldt, interview by author, 8 April 2021.
83. "Senate Amendment to HR 3338."
84. Rivard, "Collecting Disaster," 95–96.
85. Rivard, "Archiving Disaster."
86. Scheinfeldt, interview by author, 8 April 2021.
87. Brier and Brown, "The September 11 Digital Archive," 105.
88. Scheinfeldt, interview by author, 8 April 2021.
89. Pymm, "Archives and Web 2.0," 20.
90. Scheinfeldt, interview by author, 8 April 2021.
91. Brier and Brown, "The September 11 Digital Archive," 105.
92. Brier and Brown, "The September 11 Digital Archive," 106.
93. Library of Congress, "LC Accepts September 11 Digital Archive."
94. Brier and Brown, "The September 11 Digital Archive," 108.
95. Library of Congress, "Library Accepts September 11 Digital Archive, Holds Symposium."
96. Brier and Brown, "The September 11 Digital Archive," 107.
97. "September 11 Digital Archive Awarded Saving America's Treasures Grant."
98. Howard, "For Comfort and Posterity."
99. Scheinfeldt, interview by author, 8 April 2021.
100. Brett, "After 20 Years, Archive Closes."
101. Brennan and Kelly, "Why Collecting History Online Is Web 1.5."
102. "Hurricane Archive Collects over 5000 Online Stories."
103. Brennan and Kelly, "Why Collecting History Online Is Web 1.5" and Rivard, "Archiving Disaster," 4, 8, and 28.
104. Scheinfeldt, interview by author, 8 April 2021.
105. "Center for History and New Media Releases Free Software."

106. Farooqui, "Ukraine's Digital History Risking Erasure."

107. Most of the early histories of the attack centered the perspectives of everyday, direct participants. For example, see Murphy, *September 11* and Graff, *The Only Plane in the Sky*.

108. A phenomenon explored at length in Senk, "Memory Exchange," 257.

109. Senk, "Memory Exchange," 258–261.

110. Schafer et al., "Paris and Nice Terrorist Attacks," 153.

111. Jones et al., "Remembering."

Conclusion

1. Lynch, "Stewardship in the 'Age of Algorithms.' "

2. "Bibliotheca Alexandrina."

3. Choi and Jeon, "A Web Archiving System"; "Archiving Internet Information"; Lee, "Web Archiving Singapore."

4. Dwyer, "Harvesting Government History"; Paris and Currie, "How the 'Guerrilla Archivists' Saved History."

5. "Archive Dark Age Looms."

6. Friedberg, Hagler, and Land, "How E-Mail Raises the Spectre."

7. Reagan, "Digital Ice Age."

8. Bollacker, "Avoiding a Digital Dark Age."

9. Jeffrey, "A New Digital Dark Age?"

10. Sample, "Google Boss Warns."

11. Nelson, "Reactions To Vint Cerf."

12. Wernick and Intagliata, "Scientists Warn."

13. Shepherd, "Why Facebook Will Plunge Us into a Digital Dark Age."

14. Farrell et al., "Results of a Survey."

15. From extensive conversations assembled as part of my membership on the Royal Society of Canada's Task Force on Archiving COVID-19. For our findings, see Jones et al., "Remembering Is a Form of Honouring."

16. Berreby, "The Art of Losing."

17. Gillmor, "Internet's Archive Will Preserve the Web," 12.

18. Markoff, "When Big Brother Is a Librarian."

19. Mayer-Schönberger, *Delete*, 3.

20. Mayer-Schönberger, *Delete*, 52.

21. Mayer-Schönberger, *Delete*, 11.

22. A good overview in Cofone, "Online Harms."

23. Scasa, "A Little Knowledge," 38.

24. Stoddart, "Lost in Translation," 20.

25. Tretikov, "European Court Decision."

26. Dulong de Rosnay and Guadamuz, "Memory Hole," 11–12.

27. See for example Robertson, "Digitization."

28. The former culminated in my first monograph, Milligan, *Rebel Youth*.

29. Ruest and Milligan, "Open-Source Strategy."

30. Tsukayama, "CERN Reposts the World's First Web Page"; Suda, "CERN: Line Mode Browser."

31. AlSum, "Reconstruction of the US First Website."

32. Deschamps, "Future of Web Archives in Canada."
33. Deschamps, "Rare Canadian Artifacts."
34. Wikipedia, "List of Websites Founded before 1995."
35. Exploratorium, "Exploratorium Fact Sheet."
36. "Difference Engine."

Unless otherwise indicated, periodical articles with no pagination listed were found using LexisNexis Academic.

Interviews

Higgs, Edward (3 February 2021)
Kuny, Terry (24 February 2021)
Kahle, Brewster (26 February 2021)
Koerbin, Paul (1 March 2021)
Grotke, Abbie (5 March 2021)
Hedstrom, Margaret (19 March 2021)
Schneider, Steve (23 March 2021)
Foot, Kirsten (23 March 2021)
Scheinfeldt, Tom (8 April 2021)

Published Works Cited

Abbate, Janet. *Inventing the Internet.* Cambridge: MIT Press, 2000.

Adams, Margaret O., and Thomas E. Brown. "Myths and Realities about the 1960 Census." *Genealogy Notes* 32, no. 4 (Winter 2000). https://www.archives.gov/publications/prologue/2000/winter/1960-census.html#nt1

"ADL Web Site Included in Library of Congress Archive in Sept. 11 Attacks." *US Newswire,* 18 October 2001.

"AHR Forum Essay: Can We Save the Present for the Future?" *American Historical Review* 108, no. 3 (June 2003): 734.

Ainsworth, Scott G., Michael L. Nelson, and Herbert Van de Sompel. "Only One Out of Five Archived Web Pages Existed As Presented." In *Proceedings of the 26th ACM Conference on Hypertext & Social Media* (2015): 257–266.

"Alexa Internet Donates Archive of the World Wide Web to Library of Congress." *Business Wire,* 13 October 1998.

"Alexa Internet Introduces Web Navigation That Learns from People." *Business Wire,* 17 July 1997.

Alexander, Michael. "Parallel Computer Wins FLOPS Race." *Computerworld,* 25 March 1991.

Allan, Stuart. "Reweaving the Internet: Online News of September 11." In *Journalism After September 11*, edited by Barbie Zelizer and Stuart Allan, 169–190. London: Routledge, 2003.

AlSum, Ahmed. "Reconstruction of the US First Website." In *Proceedings of the 15th ACM/IEEE-CS Joint Conference on Digital Libraries* (2015): 285–286.

"Amazon.com to Pay $450-Million for Alexa, e-Niche." *National Post*, 14 May 1999.

"America Online to Buy Two Firms." *United Press International*, 22 May 1995.

"America under Attack." *CNN*, 11 September 2001. Available via the Internet Archive Wayback Machine. https://web.archive.org/web/20010911200318/http://www.cnn.com/.

Ankerson, Megan. "Take Me Back! Web History as Chronotourism of the Digital Archive." Presentation at Times and Temporalities of the Web International Symposium, Paris, France, 2015.

"A. P. Sloan Foundation Grants $700,000 to Preserve Electronic History of September 11, 2001." *PR Newswire*, 8 March 2002.

"Archive Dark Age Looms." *Sunday Territorian*, February 2, 2003.

"Archiving Internet Information." National Diet Library of Japan, accessed November 12, 2020. https://www.ndl.go.jp/en/collect/internet/index.html.

Arms, William Y. "Web Preservation Project: Final Report." Available via the Internet Archive Wayback Machine. 3 September 2001, 7. https://web.archive.org/web/20030917040530/http://www.loc.gov/minerva/webpresf.pdf.

———. "Web Preservation Project: Interim Report." Available via the Internet Archive Wayback Machine. 15 January 2001. https://web.archive.org/web/20030916234524/http://www.loc.gov/minerva/webpresi.pdf.

Arnold, John. *What Is Medieval History?* Cambridge: Polity, 2008.

Arvidson, Allan, and Krister Persson. "Kulturarw3: The Swedish WWW-Archive," slidedeck, 11 March 2001. Available via the Internet Archive Wayback Machine. https://web.archive.org/web/20030410140108/http://bibnum.bnf.fr/ecdl/2001/sweden/sld001.htm.

Aufderheide, Patricia, and Peter Jaszi. *Reclaiming Fair Use: How to Put Balance Back in Copyright*. 2nd ed. Chicago: University of Chicago Press, 2018.

Auletta, Ken. "The Microsoft Provocateur." *New Yorker*, 5 May 1997.

B., Sarah. "Internet Archive Addresses Housing Crisis with 'Foundation Housing' Project." *Richmond District Blog*, 19 March 2015. https://richmondsfblog.com/2015/03/19/internet-archive-addresses-housing-crisis-with-foundation-housing-project/.

Bak, Arpad. "Dead Media Project: An Interview with Bruce Sterling." *CTheory*. March 16, 1999. https://journals.uvic.ca/index.php/ctheory/article/view/14775/5649.

Balogun, Tolulope, and Trywell Kalusopa. "Web Archiving of Indigenous Knowledge Systems in South Africa." *Information Development* 38, no. 4 (November 2022): 658–671.

Barnes, Bill. "Nothing but Net." *Slate*, 28 February 1997. https://slate.com/technology/1997/02/nothing-but-net.html.

Barnet, Belinda. *Memory Machines: The Evolution of Hypertext*. London: Anthem Press, 2013.

Baucom, Erin. "A Brief History of Digital Preservation." In *Digital Preservation in*

Libraries: Preparing for a Sustainable Future. Edited by Jeremy Myntti and Jessalyn Zoom, 3–19. Chicago: American Library Association 2019.

Bearman, David. *Electronic Evidence: Strategies for Managing Records in Contemporary Organizations*. Pittsburgh: Archives & Museums Informatics, 1994.

Berners-Lee, Tim. *Weaving the Web: The Original Design and Ultimate Destiny of the World Wide Web*. San Francisco: HarperBusiness, 2000.

Berr, Jonathan. "Bookworms Embrace Yahoo!" *TheStreet.com*, 3 October 2005.

Berreby, David. "The Art of Losing: Why Forgetting Everything that Ever Appeared on the Web Wouldn't Be a Disaster." *Slate*, 16 July 1996. https://slate.com/news-and-politics/1996/07/the-art-of-losing.html.

Bhattacharya, Saradindu. "Mourning Becomes Electronic(a): 9/11 Online." *Journal of Creative Communications* 5, no. 1 (2010): 63–74.

"The Bibliotheca Alexandrina: A Truly Digital Library for the 21st Century." Bibliotheca Alexandrina, 20 April 2002. Available via the Internet Archive Wayback Machine. https://ia801607.us.archive.org/11/items/bibalex/bibalex_pr.pdf?cnt=0.

Billington, James H. James H. Billington to Brewster Kahle, 20 July 1994. Available at https://archive.org/details/04Kahle000826.

Bixenspan, David. "When the Internet Archive Forgets." Gizmodo, 28 November 2018. https://gizmodo.com/when-the-internet-archive-forgets-1830462131.

Blouin Jr., Francis X., and William G. Rosenberg. *Processing the Past: Contesting Authority in History and the Archives*. Oxford: Oxford University Press, 2011.

Bollacker, Kurt D. "Avoiding a Digital Dark Age." *American Scientist*, April 2010: 106.

Borsuk, Amaranth. *The Book*. Cambridge: MIT Press, 2018.

Boutin, Paul. "The Archivist." *Slate*, 7 April 2005. https://slate.com/technology/2005/04/the-internet-archive-wants-your-files.html.

Brabazon, Tara. "Dead Media: Obsolescence and Redundancy in Media History." *First Monday* 18, no. 7 (July 2013). https://firstmonday.org/ojs/index.php/fm/article/view/4466.

Brand, Stewart. *The Clock of the Long Now: Time and Responsibility*. New York: Basic, 1999.

Brennan, Sheila A., and T. Mills Kelly. "Why Collecting History Online Is Web 1.5." Center for History and New Media Case Study, March 2009. https://rrchnm.org/essay/why-collecting-history-online-is-web-1-5/.

Brett, Megan. "After 20 Years, Archive Closes to New Contributions." The September 11 Digital Archive. 23 August 2021. https://911digitalarchive.org/news/.

Brier, Stephen. "The Intentional Archive: Why Historians Need to Become Archivists (or Begin to Think and Act Like Them)." Keynote address at Choices & Challenges: Hot Topics Facing Curators and Archivists symposium, The Henry Ford Museum, Dearborn, Michigan, October 2004. Available at http://ophelia.sdsu.edu:8080/henryford_org/03-25-2015/pdf/2004/brier.pdf.

Brier, Stephen, and Joshua Brown. "The September 11 Digital Archive: Saving the Histories of September 11, 2001." *Radical History Review*, issue 111 (Fall 2011): 101–109.

Brodie, Nancy. "The National Library's Electronic Publications Pilot Project." *National Library News* 27, no. 3–4 (March/April 1995): 27–39.

Broekhuyse, Paul. "Rescuers Save Endangered Websites." *The Australian*, 29 August 2000.

Brown, Thomas E. "History of NARA's Custodial Program for Electronic Records: From the Data Archives Staff to the Center for Electronic Records, 1968–1998." In *Thirty Years of Electronic Records*, edited by Bruce I. Ambacher, 1–23. Lanham, MD: Scarecrow Press.

Brügger, Niels. *The Archived Web: Doing Web History in the Digital Age*. Cambridge: MIT Press, 2018.

Brügger, Niels, and Ian Milligan, eds. *SAGE Handbook of Web History*. London: SAGE, 2018.

Brügger, Niels, and Ralph Schroeder, eds. *The Web as History*. London: UCL Press, 2017.

Bruno, Michael P. "Archiving Anguish, Byte by Byte." *Newsbytes*, 13 September 2002.

Bruns, Axel. "Contemporary Culture and the Web." Presentation, National Library of Australia, Canberra, Australia, 9–12 November 2004. http://snurb.info/files/Contemporary%20Culture%20and%20the%20Web.ppt.

Buckland, Michael K. "Emmanuel Goldberg, Electronic Document Retrieval, and Vannevar Bush's Memex." *Journal of the American Society for Information Science* 43, no. 4 (1992): 284–294.

Buckland, Michael. *Information and Society*. Cambridge: MIT Press, 2017.

Budd, John M. "Review of Into the Future." *Library Collections, Acquisitions, and Technical Services* 23, no. 2 (1999): 225.

Burgess, Verona. "Club Buggery Joins Wiggles, Vegemite as Cultural Icons." *Canberra Times*, 2 July 1999.

Burner, Mike, and Brewster Kahle. "Arc File Format." *Internet Archive*, 15 September 1996. https://archive.org/web/researcher/ArcFileFormat.php.

Bush, Vannevar. "As We May Think." *The Atlantic*, July 1945. http://www.theatlantic.com/magazine/archive/1945/07/as-we-may-think/303881/?single_page=true.

———. "Memex Revisited." In *New Media, Old Media: A History and Theory Reader*, 2nd ed., edited by Wendy Hui Kyong Chun and Anna Watkins Fisher, with Thomas W. Keenan, 147–157. New York: Routledge, 2016.

———. *Science Is Not Enough*. New York: Morrow, 1967.

Butler, Mark. "Libraries Break Down Walls." *The Australian*, 16 February 2009.

Cannon, Angela. "The Historical Development of the Bulgarian Collection at the Library of Congress, 1894 to the Present." *Slavic & East European Information Resources* 11, no. 4 (December 2010): 305–368.

———. "The Origins of the Russian Collection at the Library of Congress (1800–1906)." *Slavic & East European Information Resources* 15, no. 1–2 (May 2014): 3–59.

Carlson, Scott, and Jeffrey R. Young. "Google will Digitize and Search Millions of Books from 5 Top Research Libraries." *Chronicle of Higher Education*, 7 January 2005.

Caruso, Denise. "Can the Web Be a Time Capsule If No One Is Saving the Information for Posterity?" *New York Times*, 11 March 1996.

———. "In a Society Dependent on Technology, Reliability Is Paramount. Problem Is, Today's Technology Isn't Reliable." *New York Times*, November 2, 1998. https://www.nytimes.com/1998/11/02/business/technology-digital-commerce-society-dependent-technology-reliability-paramount.html.

"Center for History and New Media Releases Free Software to Make Online Exhibits Easy." *US Fed News*, 21 February 2008.

Chandrasekaran, Rajiv. "In California, Creating a Web of the Past." *Washington Post*, 22 September 1996.

Chang, Chun-Chieh, Chi-Lin Chen, Chi-Wei Lin, Jen-Wei Huang, Chu-Sing Yang. "The Analysis of User Behaviour and Relationship on Bulletin Board Systems." *Conference on Technologies and Applications of Artificial Intelligence*, November 2015.

Choi, Kyung Ho, and Dal Ju Jeon. "A Web Archiving System of the National Library of Korea: OASIS." *Digital Libraries: Achievements, Challenges and Opportunities*, edited by Shigeo Sugimoto, Jane Hunter, Andreas Rauber, and Atsuyuki Morishima, 313–322. Lecture Notes in Computer Science. Berlin, Heidelberg: Springer, 2006.

Christian, Jon. "Why Does Amazon Have Two Completely Different Products Called Alexa?" *The Outline*, 6 October 2017. https://theoutline.com/post/2377/why-does-amazon-have-two-completely-different-products-called-alexa.

Chun, Wendy Hui Kyong. *Programmed Visions: Software and Memory*. Cambridge: MIT Press, 2013.

Churbuck, David C. "Good-Bye, Dewey Decimals." *Forbes*, 15 February 1993.

Clark, Jim. *Netscape Time: The Making of the Billion-Dollar Start-Up that Took on Microsoft*. New York: St. Martin's, 1999.

Cleveland, Gary. "Overview of Document Management Technology." *International Federation of Library Associations and Institutions*, June 1995. https://archive.ifla.org/VI/5/op/udtop2/udt-op2.pdf.

Cofone, Ignacio N. "Online Harms and the Right to Be Forgotten." In *The Right to Be Forgotten: A Canadian and Comparative Perspective*, edited by Ignacio N. Cofone, 1–16. New York: Routledge, 2020.

"The Constitution of the United States: A Transcription." National Archives of the United States. https://www.archives.gov/founding-docs/constitution-transcript.

Cook, Terry. "The Archive(s) Is a Foreign Country: Historians, Archivists, and the Changing Archival Landscape." *Canadian Historical Review* 90, no. 3 (September 2009): 497–534.

———. "Easy to Byte, Harder to Chew: The Second Generation of Electronic Record Archives." *Archivaria* 33 (Winter 1991–92): 202–216.

"Copyright Act 1968—Sect 200AB: Use of Works and Other Subject-Matter for Certain Purposes." *Commonwealth Consolidated Acts*. http://www6.austlii.edu.au/cgi-bin/viewdoc/au/legis/cth/consol_act/ca1968133/s200ab.html.

"Copyright Term Extension." Public Law 105-298, 105th Congress, 27 October 1998. https://www.govinfo.gov/content/pkg/PLAW-105publ298/pdf/PLAW-105publ298.pdf.

Corrado, Edward M., and Heather Lea Moulaison, eds. *Digital Preservation for Libraries, Archives, & Museums*. Lanham: Rowman & Littlefield, 2014.

Cox, Richard J. *The First Generation of Electronic Records Archivists in the United States*. New York: Haworth Press, 1994.

Cox, Richard J., Mark K. Biagini, Toni Carbo, Tony Debons, Ellen Detlefsen, Jose Mare Griffiths, Don King, David Robins, Richard Thompson, Chris Tomer, and Martin Weiss. "The Day the World Changed: Implications for Archival, Library, and Information Science Education." *First Monday* 6, no. 12 (December 2001). https://firstmonday.org/ojs/index.php/fm/article/view/908.

"Cracking Wise on the Internet." *Ottawa Citizen*, 16 November 1996, B8.
Cramb, Alex. "Cyber Historian." *Communique*, 13 October 1997. As found in https://archive.org/details/alexainternetpre19unse_1.
Crew, Spencer R. Spencer R. Crew to Alan Keyes (Presidential Candidate), 28 February 1996. Available at https://archive.org/details/04Kahle000461.
Crockatt, Richard. *America Embattled: September 11, Anti-Americanism and the Global Order*. London: Routledge, 2003.
Crymble, Adam. *Technology and the Historian: Transformations in the Digital Age*. Urbana: University of Illinois Press, 2021.
Cunningham, Michael. "Brewster's Millions," *Irish Times* (Dublin, Ireland). 27 January 1997.
———. "Who Archives the World Wide Web?" *Irish Times* (Dublin, Ireland), 27 January 1997.
Davis, Ben H., and Margaret MacLean. "Mapping the Project, Grasping the Consequences." In *Time and Bits: Managing Digital Continuity*, edited by Margaret MacLean and Ben H. Davis, 1–10. Los Angeles: J. Paul Getty Trust, 1998.
De Kosnik, Abigail. *Rogue Archives: Digital Cultural Memory and Media Fandom*. Cambridge: MIT Press, 2016.
Dear, Brian. *The Friendly Orange Glow: The Untold Story of the Rise of Cyberculture*. New York: Vintage, 2018.
"Decoding Common Internet Error Messages." *Newsbytes*, 21 November 1995.
Deger, Renee. "Internet Historian Creates Tools to Inform Web Site Visitors." *ZDNet*, 18 July 1997.
Delsey, Tom. "The National Library's Role in Facilitating Scholarly Communications." *Canadian Journal of Communication* 22, no. 3 (1997): 89.
Deschamps, Ryan. "The Future of Web Archives in Canada: The Inaugural Meeting of the Canadian Web Archives Coalition." *Archives Unleashed News Blog*, September 28, 2017. https://news.archivesunleashed.org/the-future-of-web-archives-in-canada-the-inaugural-meeting-of-the-canadian-web-archives-coalition-83872b114fb3.
———. "Rare Canadian Artifacts from the Early Web." *Archives Unleashed News Blog*, August 2, 2018. https://news.archivesunleashed.org/rare-canadian-artifacts-from-the-early-web-652caab1d9c1.
Dick, Jeff. "Into the Future: On the Preservation of Knowledge in the Electronic Age." *Booklist* 94, no. 15 (April 1998): 1336.
Dickins, Jim. "Library's Porn Plan Opposed." *Herald Sun*, 27 August 2002.
"Difference Engine: Lost in Cyberspace." *The Economist*, September 1, 2012. http://www.economist.com/node/21560992.
Ding, Jennifer. "Can Data Die? Why One of the Internet's Oldest Images Lives On Without Its Subject's Consent." *The Pudding*, October 2021. https://pudding.cool/2021/10/lenna/.
Documenting the Digital Dark Age. "Documenting the Digital Age: Overview." Archived February 1997. Available via the Internet Archive Wayback Machine. https://web.archive.org/web/19980626155302/http://www.dtda.com/overview/index.htm.
Driscoll, Kevin. *The Modem World: A Prehistory of Social Media*. New Haven: Yale University Press, 2022.
Duffy, Clare, and Kerry Flynn. "Some of the Most Iconic 9/11 News Coverage Is Lost.

Blame Adobe Flash." *CNN.com*, 10 September 2021. https://www.cnn.com/2021/09/10/tech/digital-news-coverage-9-11/index.html.

Dulong de Rosnay, Melanie, and Andres Guadamuz. "Memory Hole or Right to Delist? Implications of the Right to Be Forgotten for Web Archiving." *RESET: Recherches en sciences sociales sur internet*, no. 6 (2017): 1–22.

Dunleavey, M. P. "Going Giddy Just Thinking of a Windfall." *New York Times*, 14 July 2007.

Dwyer, Jim. "Harvesting Government History, One Web Page at a Time." *New York Times*, December 2, 2016.

Easton, Stephen. "Australia's Top Librarian Tells How the National Library Fosters a Culture of In-House Innovation. In Two Words: 'Radical Incrementalism.'" *The Mandarin*, 24 June 2019. https://www.themandarin.com.au/110303-australias-top-librarian-tells-how-the-national-library-fosters-a-culture-of-in-house-innovation-in-two-words-radical-incrementalism/.

"Editorial." *History and Computing* 4, no. 3 (1992): ii.

"Election 2000, as It Happened." Library of Congress Information Bulletin, July/August 2001. Available via the Library of Congress WayBack Machine. https://webarchive.loc.gov/all/20010907220254/http:/www.loc.gov/loc/lcib/01078/archiving.html#loclr=blogsig.

"Election 2000 Collection." Archived 11 November 2001. Available via the Internet Archive Wayback Machine. https://web.archive.org/web/20011111021222/http://archive0.alexa.com:80/collections/e2k.html.

Electronic Publications Pilot Project Team and Electronic Collections Committee (EPPP), *Electronic Publications Pilot Project (EPPP)*. National Library of Canada, June 1996. https://www.nlc-bnc.ca/obj/p4/f2/e-report.pdf.

Emerson, Lori. *Reading Writing Interfaces: From the Digital to the Bookbound*. Minneapolis: University of Minnesota Press, 2014.

Ernst, Wolfgang. "Media Archeography: Method and Machine versus History and Narrative of Media." In *Media Archaeology: Approaches, Applications, and Implications*, edited by Erkki Huhtamo and Jussi Parikka, 239–255. Berkeley: University of California Press, 2011.

Exploratorium. "Exploratorium Fact Sheet." July 7, 2014. https://www.exploratorium.edu/about/fact-sheet.

Farooqui, Salmaan. "With Ukraine's Digital History Risking Erasure, Canadians Join Efforts to Preserve It," *Globe and Mail*. 11 April 2022. https://www.theglobeandmail.com/canada/article-canadians-join-global-efforts-to-save-ukraines-digital-archives/.

Farrell, Matthew, Edward McCain, Maria Praetzellis, Grace Thomas, and Paige Walker. "Results of a Survey of Organizations Preserving Web Content," National Digital Stewardship Alliance, October 2018. https://osf.io/ht6ay/.

"FED: Foreign Affairs Website Named as Site of Significance." *AAP Newsfeed*. 3 July 1999.

Feijen, Martin. "DNEP: The Dutch Repository of Electronic Publications." *Program* 30, no. 2 (January 1, 1996): 149–156.

"Film on Preservation in the Electronic Age Available." *Information Technology and Libraries* 16, no. 3 (September 1997): 144.

Fishbein, Meyer H. "Appraising Information in Machine Language Form." *American Archivist* 35, no. 1 (January 1972): 35–43.

———. "Welcome." In *The National Archives and Statistical Research*, edited by Meyer H. Fishbein, 3–5. Athens: Ohio University Press, 1973.

Fleckner, John. "F. Gerald Ham: Jeremiah to the Profession." *The American Archivist* 77, no. 2 (October 1, 2014): 377–393.

Flynn, Laurie J. "Alexa Internet: The Search as a Communal Effort." *New York Times*, 19 July 1997.

———. "A New Power in Publishing." *New York Times*, 24 April 1994.

Foot, Kirsten A., and Steven M. Schneider. *Web Campaigning*. Cambridge, MA: MIT Press, 2006.

Foster, Andrea L. "2 Scholars Archive Web Sites on Terrorist Attacks." *Chronicle of Higher Education*, 18 September 2001. https://www.chronicle.com/article/2-scholars-archive-web-sites-on-terrorist-attacks/.

Foucault, Michel. *The Archaeology of Knowledge*. Routledge Classics Edition. London: Routledge, 2002.

Friedberg, Errol C., Herbert K. Hagler, and Kevin J. Land. "How E-Mail Raises the Spectre of a Digital Dark Age." *Nature* 423, no. 801 (June 19, 2003): 801.

Friedrich, Markus. "Archivists." In *Information: A Historical Companion*, edited by Ann Blair, Paul Duguid, Anja-Silvia Goeing, and Anthony Grafton, 312–317. Princeton, NJ: Princeton University Press, 2021.

———. *The Birth of the Archive*. Translated by John Noël Dillon. Ann Arbor: University of Michigan Press, 2018.

Gabriele, Matthew, and David M. Perry. *The Bright Ages: A New History of Medieval Europe*. New York: HarperCollins, 2022.

Gaffield, Chad. "Clio and Computers in Canada and Beyond: Reflections on the Contested Past and Possible Future for Digital History." *Canadian Historical Review*, 101, no. 4 (December 2020): 559–584.

Gardner, James B. "Report on Documenting the Digital Dark Age," Final Report Prepared for the National Science Foundation, 1997.

Gardner, James B., and Sarah M. Henry. "September 11 and the Mourning After: Reflections on Collecting and Interpreting the History of Tragedy." *Public Historian* 24, no. 3 (Summer 2002): 37–52.

Garreau, Joel. "A Shaken Global Village on the Internet." *Washington Post*, 12 September 2001.

Garsson, Robert M. "Artificial Intelligence: Some Believe It's the Financial Expert of the Future; But Others Voice Doubts About Its Successful Long-Range Application," *American Banker*, 15 October 1984.

Geary, Patrick J. *Phantoms of Remembrance: Memory and Oblivion at the End of the First Millennium*. Princeton: Princeton University Press, 1994.

Gillmor, Dan. "Internet's Archive Will Preserve the Web for Posterity." *St. Louis Post-Dispatch*, 9 September 1996, 12.

Gitelman, Lisa. *Always Already New: Media, History, and the Data of Culture*. Cambridge: MIT Press, 2008.

Goel, Vinay. "Beta Wayback Machine—Now with Site Search!" Internet Archive Blogs, 24 October 2016. http://blog.archive.org/2016/10/24/beta-wayback-machine-now-with-site-search/.

Goggin, Gerard, and Mark McLelland. "Internationalizing Internet Studies: Beyond

Anglophone Paradigms." In *Internationalizing Internet Studies: Beyond Anglophone Paradigms*, edited by Gerard Goggin and Mark McLelland, 3–17. London: Routledge, 2009.
———. "Introduction: Global Coordinates of Internet Histories." In "Histories," in *Routledge Companion to Global Internet Histories*, edited by Gerard Goggin and Mark McLelland, 1–19. London: Routledge, 2017.
Gomes, Daniel, Elena Demidova, Jane Winters, and Thomas Risse, eds. *The Past Web: Exploring Web Archives*. Cham: Switzerland, Springer, 2021.
Government of Canada. Library and Archives of Canada Act. S.C. 2004 c. 11. Ottawa, 2004. https://laws-lois.justice.gc.ca/eng/acts/L-7.7/page-1.html#h-345269.
Government of Sweden. *Tryckfrihetsförordningen*. Issued 1949, modified 2018. https://www.riksdagen.se/sv/dokument-lagar/dokument/svensk-forfattningssamling/tryckfrihetsforordning-1949105_sfs-1949-105.
Graff, Garrett M. *The Only Plane in the Sky: An Oral History of 9/11*. New York: Avid Reader, 2019.
Gray, Douglas F. "Amazon Unit Faces $1.9 Million Payout in Privacy Settlement." *InfoWorld Daily News*, 27 April 2001.
Grose, Simon. "National Library Stocktaking." *Canberra Times*, 20 June 2005.
———. "Staff Harvest Web for History." *Canberra Times*, 28 March 2005.
Grotke, Abbie. "Ask the Recommending Officer: The September 11, 2001 Web Archive." *The Signal* (Blog), 9 September 2011. https://blogs.loc.gov/thesignal/2011/09/ask-the-recommending-officer-the-september-11-2001-web-archive/.
Hafner, Katie. "In Challenge to Google, Yahoo Will Scan Books." *New York Times*, 3 October 2005.
———. "Microsoft to Offer Online Book-Content Searches." *New York Times*, 26 October 2005.
———. "Saving the Nation's Digital Legacy." *New York Times*, 27 July 2000.
Ham, F. Gerald. "The Archival Edge." *American Archivist* 38, no. 1 (January 1975): 5–13.
Harmon, Amy. "The Talk Online; Web Offers Both News and Comfort." *New York Times*, 12 September 2001.
Hedstrom, Margaret. *Archives & Manuscripts: Machine-Readable Records*. Chicago: Society of American Archivists, 1984.
———. "Digital Preservation: A Time Bomb for Digital Libraries." *Computers and the Humanities* 31, no. 3 (1997): 189–202.
———. "Electronic Archives: Integrity and Access." In *Networking in the Humanities*, edited by Stephanie Kenna and Seamus Ross, 77–95. London: Bowker Saur, 1995.
———. "How Do We Make Electronic Archives Usable and Accessible?" *DTDA Webpage*. Available via the Internet Archive Wayback Machine. https://web.archive.org/web/19980626160605/http://www.dtda.com/presenta/hedst01.htm.
Heffernan, Virginia. *Magic and Loss: The Internet as Art*. New York: Simon & Schuster, 2016.
Hegarty, Kieran. "The Invention of the Archived Web: Tracing the Influence of Library Frameworks on Web Archiving Infrastructure." *Internet Histories* 6, no. 4 (2022): 1–20.
Helm, Leslie. "Amazon to Buy 3 More Internet Firms." *Los Angeles Times*, 27 April 1999.

Herres, Örjan. "Project Kulturarw3 ur ett open source perspektiv," Masters Thesis, Uppsala University, 2003.

Hickey, John. "The Wayback Machine: Fighting Digital Extinction in New Ways." Internet Archive Blogs, 18 October 2019. http://blog.archive.org/2019/10/18/the-wayback-machine-fighting-digital-extinction-in-new-ways/.

Higgs, Edward. "Historians, Archivists, and Electronic Recordkeeping in British Government." In *History and Electronic Artefacts*, edited by Edward Higgs, 133–151. Oxford: Clarendon Press, 1998.

Hillis, Danny. "The Millennium Clock," *Wired*, 6 December 1995. https://www.wired.com/1995/12/the-millennium-clock/.

Hirtle, Peter B. "The History and Current State of Digital Preservation in the United States." In *Metadata and Digital Collections: A Festschrift in Honor of Tom Turner*, 121–140. Ithaca: Cornell University Library Initiatives in Publishing, 2008. https://ecommons.cornell.edu/handle/1813/45862.

Hodges, Doug, and Carrol D. Lunau, "The National Library of Canada's Digital Library Initiatives." *Library Hi Tech* 17, no. 2 (January 1, 1999): 152–164.

Hollier, Anita. "The Archivist in the Electronic Age." *High Energy Physics Libraries Webzine*, issue 3 (March 2001). https://webzine.web.cern.ch/3/papers/5/.

Howard, Jennifer. "For Comfort and Posterity, Digital Archives Gather Crowds," *Chronicle of Higher Education*, 18 November 2013.

Huhtamo, Erkki, and Jussi Parikka. "Introduction." In *Media Archaeology: Approaches, Applications, and Implications*, edited by Erkki Huhtamo and Jussi Parikka, 1–21. Berkeley: University of California Press, 2011.

"Hurricane Archive Collects over 5000 Online Stories and Images; Preserving the Past through Digital Memory Banks Is Growing Trend." *US Newswire*, 25 August 2006.

"Idea Hatchery Colloquium." 6 August 1996. Available via the Internet Archive Wayback Machine. https://web.archive.org/web/20021218220811/http://www.kb.se/kw3/1996-08-06.html.

International Internet Preservation Consortium. "About the Consortium." 3 June 2004. Available via the Internet Archive Wayback Machine. https://web.archive.org/web/20040603014115/http://netpreserve.org/about/index.php.

"Internet Library Enables Users to Surf the Web's Past." *Business Wire*, 18 June 2001.

Jackson, Tim. "Archive Holds Wealth of Data." *Financial Times*, 24 November 1997.

Jeffrey, Stuart. "A New Digital Dark Age? Collaborative Web Tools, Social Media and Long-Term Preservation." *World Archaeology* 44, no. 4 (December 1, 2012): 553–570.

Johns, Adrian. *Piracy: The Intellectual Property Wars from Gutenberg to Gates*. Chicago: University of Chicago Press, 2009.

Jones, Esyllt W., Shelley Sweeney, Ian Milligan, Greg Bak, and Jo-Anne McCutcheon. "Remembering Is a Form of Honouring: Preserving the COVID-19 Archival Record." *FACETS*, 6, no. 1 (April 2021): 545–568.

Kahle, Brewster. "Announcing the Open Content Alliance." *Yahoo! Search Blog*, 2 October 2005. Available via Internet Archive Wayback Machine at https://web.archive.org/web/20051124051631/http:/www.ysearchblog.com/archives/000192.html.

———. "Archiving the Internet." *Scientific American*, April 11, 1996. https://web.archive.org/web/20060702081900/http://www.uibk.ac.at/voeb/texte/kahle.html.

———. "Archiving the Internet: Towards a Core Internet Service." 26 April 1996. https://archive.org/details/04Kahle002583.

———. "Archiving the Internet." *DTDA Webpage.* Archived 26 June 1998. Available via the Internet Archive Wayback Machine. https://web.archive.org/web/19980626160605/http://www.dtda.com/presenta/kahle01.htm.

———. "A Book Grab by Google." *Washington Post,* 19 May 2009.

———. Brewster Kahle to WAIS Employees, 18 June 1994. Available at https://archive.org/details/wais.

———. Fax from Brewster Kahle to David Cole, 11 January 1995. Available at https://archive.org/details/09Kahle000639.

———. "The Internet Companion Company." 19 June 1996. Available at https://archive.org/details/03Kahle001132.

———. "Universal Access to All Knowledge." *American Archivist,* 70, no. 1 (2007): 23–31.

Katz, Stan. "Publishing History Digitally." *Brainstorm Blog,* 11 May 2010. Available via the Internet Archive Wayback Machine. https://web.archive.org/web/20170824153530/https://chronicle.com/blogs/brainstorm/publishing-history-digitally/23894.

Kelly, Kevin. "Scan This Book!" *New York Times Magazine,* 14 May 2006.

Kenney, Anne, and Lynne K. Personius. *The Cornell/Xerox Commission on Preservation and Access Joint Study in Digital Preservation. Report Phase 1* (January 1990–December 1991. Commission on Preservation and Access, September 1992. https://files.eric.ed.gov/fulltext/ED352040.pdf.

Kenny, Anthony. "Keynote Address," In *Networking in the Humanities,* edited by Stephanie Kenna and Seamus Ross, 1–13. London: Bowker Saur, 1995.

Kesner, Richard M. "Automated Information Management: Is There a Role for the Archivist in the Office of the Future?" *Archivaria* 19 (Winter 1984–85): 162–172.

Kimpton, Michele, and Jeff Ubois. "Year-by-Year: From an Archive of the Internet to an Archive on the Internet." In *Web Archiving,* edited by Julien Masanès, 201–212. Berlin and Heidelberg: Springer, 2006.

Kirschenbaum, Matthew G. *Bitstreams: The Future of Digital Literary Heritage.* Philadelphia: University of Pennsylvania Press, 2021.

———. *Mechanisms: New Media and the Forensic Imagination.* Cambridge: MIT Press, 2008.

Koerbin, Paul. "National Web Archiving in Australia: Representing the Comprehensive." In *The Past Web: Exploring Web* Archives, edited by Daniel Gomes, Elena Demidova, Jane Winters, and Thomas Risse, 23–32. Berlin: Springer, 2021.

Koman, Richard. "Riding Along with the Internet Bookmobile." *Salon.com,* 9 October 2002. https://www.salon.com/2002/10/09/bookmobile/.

Kornblum, Janet. "Web-Page Database Goes Wayback When." *USA Today,* 30 October 2001.

Koster, Martijn. "ANNOUNCE: A Standard for Robot Exclusion." 3 July 1994, posted to WWW-Talk. Available at http://ksi.cpsc.ucalgary.ca/archives/WWW-TALK/www-talk-1994q3/0007.html.

Kresh, Diane Nester. "Courting Disaster: Building a Collection to Chronicle 9/11 and Its Aftermath." *Information Bulletin,* September 2002.

Kuny, Terry. "A Digital Dark Ages? Challenges in the Preservation of Electronic Information." Paper presented at the 63rd IFLA Council and General Conference, 27 August 1997. http://archive.ifla.org/IV/ifla63/63kuny1.pdf.

Kurutuz, Steven. "Meet Your Meme Lords." *New York Times*, 7 April 2020. https://www.nytimes.com/2020/04/07/style/internet-archive-library-congress.html.

Lane, Carole A. *Naked in Cyberspace: How to Find Personal Information Online*. Medford, NJ: Information Today, 2002.

Lee, Ivy. "Web Archiving Singapore." International Preservation Consortium Annual Meeting. Zagreb, Croatia, June 2019. https://netpreserve.org/ga2019/wp-content/uploads/2019/07/IIPCWAC2019-IVY_LEE-Sharing_by_the_National_Library_Singapore_on_the_journey_towards_collecting_digital_malterials.pdf.

Lemercier, Claire, and Claire Zalc, *Quantitative Methods in the Humanities: An Introduction*, trans. Arthur Goldhammer. Virginia: University of Virginia Press, 2019.

Lesk, Michael. "Foreword." In Edward M. Corrado and Heather Lea Moulaison, *Digital Preservation for Libraries, Archives, & Museums*. Lanham: Rowman & Littlefield, 2014: xvii–xx.

———. "Preservation of New Technology: A Report of the Technology Assessment Advisory Committee to the Commission on Preservation and Access." Commission on Preservation and Access, Washington DC. October 1992.

Lessig, Lawrence. *Free Culture: How Big Media Uses Technology and the Law to Lock Down Culture and Control Creativity*. London: Penguin Press, 2004

Library of Congress. "LC Accepts September 11 Digital Archive." *Computers in Libraries* 23, no. 10 (November 2003), 52.

———. "Library Accepts September 11 Digital Archive, Holds Symposium." 15 August 2003. https://www.loc.gov/item/prn-03-142/library-accepts-september-11-digital-archive-holds-symposiu/2003-08-15/.

Lidman, Tomas. "Ny förordning för Kulturarvsprojektet." *Kungliga biblioteket*. 5 June 2002. Available via the Internet Archive Wayback Machine. https://web.archive.org/web/20021214135826/https://www.kb.se/Info/Pressmed/Arkiv/2002/020605.htm.

Lilleniit, Roselyn. "The National Library of Canada's Electronic Publications Pilot Project." *The Serials Librarian* 27, no. 4 (April 10, 1996): 27–39. https://doi.org/10.1300/J123v27n04_03.

Lin, Jimmy, et al. "We Could, but Should We? Ethical Considerations for Providing Access to GeoCities and Other Historical Digital Collections." In *Proceedings of the 2020 Conference on Human Information Interaction and Retrieval* (2020): 135–144.

Livingston, Jessica. *Founders at Work: Stories of Startups' Early Days*. New York: Apress, 2008.

Lomborg, Stine. "Ethical Considerations for Web Archives and Web History Research." In *SAGE Handbook of Web History*, edited by Neils Brügger and Ian Milligan, 99–111. London: SAGE Publications, 2018.

———. "Personal Internet Archives and Ethics." *Research Ethics* 9, no. 1. (March 1, 2013): 20–31. https://doi.org/10.1177/1747016112459450.

Lor, Peter, and Johannes J. Britz. "A Moral Perspective on South-North Web Archiving." *Journal of Information Science* 30, no. 6 (2004): 540–549.

Lovejoy, Margot. "Artists' Books in the Digital Age." *SubStance* 26, no. 1 (1997): 113–134.

Lukanuski, Mary. "Help Is on the WAIS." *American Libraries*, October 1992.

Lunn, Jean. "The National Library of Canada, 1950–1968." *Archivaria* 15 (January 1982): 86–95.

Lyman, Peter, and Howard Besser. "Defining the Problem of Our Vanishing Memory: Background, Current Status, Models for Resolution." In *Time and Bits: Managing Digital Continuity*, edited by Margaret MacLean and Ben H. Davis, 11–20. Los Angeles: J. Paul Getty Trust, 1998.

Lynch, Clifford. "Stewardship in the 'Age of Algorithms.'" *First Monday* 22, no. 12 (2017). https://doi.org/10.5210/fm.v22i12.8097.

Lyons, Susan. "Preserving Electronic Government Information," *Reference Librarian* 49, no. 94 (2006): 207–223.

MacLean, Margaret. "Setting the Stage: Summary of Initial Discussions." In *Time and Bits: Managing Digital Continuity*, edited by Margaret MacLean and Ben H. Davis, 32–33. Los Angeles: J. Paul Getty Trust, 1998.

MacLean, Margaret, and Ben H. Davis, eds. *Time & Bits: Managing Digital Continuity*. Los Angeles: J. Paul Getty Trust, 1998.

Maemura, Emily, Nicholas Worby, Ian Milligan, and Christoph Becker. "If These Crawls Could Talk: Studying and Documenting Web Archives Provenance." *Journal of the Association for Information Science and Technology* 69, issue 10 (October 2018): 1223–1233.

Malloy, Judy. "The Origins of Social Media." In *Social Media Archeology and Poetics*, edited by Judy Malloy, 3–50. Cambridge: MIT Press, 2016.

Mannerheim, Johan. "Preserving the Digital Heritage of the World: Some Thoughts after Having Collected 30 Million Swedish Web Pages." *Human IT* 4, no. 1 (2000). https://humanit.hb.se/article/view/193.

Manning, Ric. "Alexa Brings Dead Webs Back to Life," *Manning the Wires*, February 1998.

Marcus, John. "US Starts Archive of Whole Web." *Times Higher Education*, 13 December 1996.

Markoff, John. "Data-Gathering Program Ignites Technologists' Battle Over Privacy." *San Diego Union-Tribune*, 11 January 2000.

———. "For Shakespeare, Just Log On." *New York Times*, 3 July 1991.

———. "New Service Tracks Web Use." *New York Times*, 21 July 1997.

———. "Supercomputer Pioneer Seeks Investor or Buyer." *New York Times*, 28 July 1994.

———. "When Big Brother Is a Librarian." *New York Times*, 9 March 1997.

Martin, Lauren. "Where Websites Go to Die." *Sydney Morning Herald*, 17 October 2003.

Masanès, Julien. "Selection for Web Archives." In *Web Archiving*, edited by Julien Masanès, 71–92. Heidelberg; Berlin: Springer, 2006.

———. "Web Archiving: Issues and Methods." In *Web Archiving*, edited by Julien Masanès, 1–53. Berlin: Springer, 2006.

Mayer-Schönberger, Viktor. *Delete: The Virtue of Forgetting in the Digital Age*. Princeton: Princeton University Press, 2009.

Mayfield, Kendra. "How to Preserve Digital Art." *Wired*, 23 June 2002. https://www.wired.com/2002/07/how-to-preserve-digital-art/.

McKenzie, Matt. "Alexa Gets You Where You Want to Go." *Seybold Report on Internet Publishing*, August 1997.

McLaughlin, Margaret L. "The Art Site on the World Wide Web." *Journal of Computer-Mediated Communication* 1, no. 4 (March 1996): 51–79.

McNeil, Joanne. *Lurking: How a Person Became a User.* New York: Farrar, Straus and Giroux, 2020.

Meacham, Steve. "Editing Our future." *Sydney Morning Herald,* 4 May 2005.

———. "Open All Hours." *Sydney Morning Herald,* 18 April 2004.

Meddis, Sam Vincent. "Lots of Good Things and Turbulent Times." *USA Today,* 19 July 2001.

Mieszkowski, Katharine. "Dumpster Diving on the Web." *Salon.com,* 2 November 2001. https://www.salon.com/2001/11/02/wayback/.

Miller, Leslie. "You are a Database and Access Abounds," *USA Today,* 9 June 1997.

Milligan, Ian. *History in the Age of Abundance? How the Web Is Transforming Historical Research.* Kingston and Montreal: McGill–Queen's University Press, 2019.

———. *Rebel Youth: 1960s Labour Unrest, Young Workers, and New Leftists in English Canada.* Vancouver: UBC Press, 2014.

———. *The Transformation of Historical Research in the Digital Age.* Cambridge: Cambridge University Press, 2022.

Milligan, Ian, and Thomas J. Smyth. "Studying the Web in the Shadow of Uncle Sam: The Case of the .ca Domain." In *The Historical Web and Digital Humanities,* edited by Niels Brügger and Ditte Laursen, 45–63. New York: Routledge, 2019.

Mirapaul, Matthew. "How the Net Is Documenting a Watershed Moment." *New York Times,* 15 October 2001.

Moozakis, Chuck. "The Daunting Task of Storing the Web." *InternetWeek,* 13 October 1997.

Morris, R. J. "Electronic Documents and the History of the Late 20th Century Black Holes or Warehouses—What Do Historians Really Want?" In *Electronic Information Resources and Historians: European Perspectives,* edited by Seamus Ross and Edward Higgs, 302–316. St. Katharinen: Max-Planck-Institut für Geschichte in Kommission bei Scripta Mercaturae Verlag, 1993.

Morrison, Ian. "Www.Nla.Gov.Au/Pandora: Australia's Internet Archive." *Australian Library Journal* 48, no. 3 (January 1, 1999): 271–284.

Murphy, Dean E. *September 11: An Oral History.* New York: Doubleday, 2002.

Murphy, Leila. Fax from Leila Murphy to Brewster Kahle, 4 March 1996. Available at https://archive.org/details/04Kahle000476.

Myhrvold, Nathan. "Internet History Conference—Position Paper." February 1997. In Margaret Hedstrom Personal Collection.

———. "Road Kill on the Information Highway." InterOffice Microsoft Memo. 8 September 1993. https://sriramk.com/memos/nathan-roadkill.pdf.

———. "Why Archive the Internet?" *DTDA Webpage.* Archived 26 June 1998. Available via the Internet Archive Wayback Machine. https://web.archive.org/web/19980626160605/http://www.dtda.com/presenta/myhrv001.htm.

National Historical Publications and Records Commission (NHPRC). "Research Issues in Electronic Records." Washington DC: National Historical Publications and Records Commission, 1991.

National Library of Australia. "Guidelines for the Selection of Online Australian Publications Intended for Preservation." Last modified 21 December 1999. Available via

the Internet Archive Wayback Machine. https://web.archive.org/web/20000709 214101/http://www.nla.gov.au/scoap/guidelines.html.
———. "National Strategy for Provision of Access to Australian Electronic Publications: A National Library of Australia Position Paper." 24 December 1996. https://web archive.nla.gov.au/awa/19970212064854/http://www.nla.gov.au/policy/paep.html.
———. "Selection Committee on Online Australian Publications." 12 February 1997. https://webarchive.nla.gov.au/awa/19970212062836/http://www.nla.gov.au/1 /scoap/scoapgui.html.
———. "What We Collect: Ephemera." 2021. https://www.nla.gov.au/what-we-collect /ephemera.
"National Museum of American History Sets Up Katrina Collection; Collection to Document Unprecedented Historical Events in Gulf Coast Region." *US Newswire*, 2 December 2005.
"National Museum of American History to Commemorate 10th Anniversary of Sept. 11 Attacks." *Targeted News Service*, 16 June 2011.
National Public Radio, "Problems of Digital Preservation Show Transcript with Ira Flatow and Brewster Kahle." Talk of the Nation, 27 April 2001.
National Research Council. *LC21: A Digital Strategy for the Library of Congress*. Washington, DC: The National Academies Press, 2000. https://doi.org/10.17226/9940.
National Research Council of the National Academies. *The Internet Under Crisis Conditions*. Washington: National Academies Press, 2003.
Nelson, Michael L. "Reactions To Vint Cerf's 'Digital Vellum.'" *Web Science and Digital Libraries Research Group* (blog), February 17, 2015. https://ws-dl.blogspot.com/2015 /02/2015-02-17-reactions-to-vint-cerfs.html.
Nelson, T. H. "Complex Information Processing: A File Structure for the Complex, the Changing and the Indeterminate." In *Proceedings of the 1965 20th National ACM Conference* (1965): 84–100.
Nelson, Theodor Holm. "Xanalogical Structure, Needed Now More than Ever: Parallel Documents, Deep Links to Content, Deep Versioning, and Deep Re-Use." *ACM Computing Surveys* 31, no. 4es (December 1, 1999): 33–es.
Nilsen, Kristi. "Government Information Policy in Canada." *Government Information Quarterly* 11, no. 2 (January 1, 1994): 191–209.
"911history.net." Archived 21 January 2002. Available via the Internet Archive Wayback Machine. https://web.archive.org/web/20020121155416/http://www.911history.net:80/.
Nordic Web Archive. "About the NWA Toolset." Last modified 5 September 2002. Available via the Internet Archive Wayback Machine. https://web.archive.org/web /20021211215815/http://nwa.nb.no/aboutNwaT.php.
Notess, Greg R. "The Wayback Machine: The Web's Archive." *Online* 26, no. 2 (March 2002): 59–61.
Novick, Peter. *That Noble Dream: The "Objectivity Question" and the American Historical Profession*. 1st ed. Cambridge: Cambridge University Press, 1988.
Nowviskie, Bethany. "Digital Humanities in the Anthropocene." *Digital Scholarship in the Humanities* 30, issue supplement 1 (December 2015): i4–i15.
"The Oakland Archival Policy: Recommendations for Managing Removal Requests and Preserving Archival Integrity." 13–14 December 2002. Available at https://groups .ischool.berkeley.edu/archive/aps/removal-policy.

O'Brien, Chris. "He Fights for Open Access to the World's Digital Library," *San Jose Mercury News*, 30 October 2005.

O'Brien, Danny. "Online Librarian Plans Giveaway with No Returns." *Irish Times*, 13 December 2002.

Ogden, Jessica. "Saving the Web: Facets of Web Archiving in Everyday Practice." Unpublished doctoral dissertation. University of Southampton, 2020.

Ogden, Sherelyn. "Review of Into the Future." *Library Quarterly* 70, no. 1 (Jan 2000): 145.

"One Year Later: September 11 and the Internet." Pew Internet and American Life Project, 5 September 2002. https://www.pewresearch.org/internet/2002/09/05/one-year-later-september-11-and-the-internet/.

Openness Advisory Panel. "Responsible Openness: An Imperative for the Department of Energy." Secretary of Energy Advisory Board. 25 August 1997. https://sgp.fas.org/advisory/oaprept.html.

Ovenden, Richard. *Burning the Books: A History of Knowledge Under Attack*. London: John Murray Press, 2020.

Owens, Trevor. *The Theory and Craft of Digital Preservation*. Baltimore: Johns Hopkins University Press, 2018.

Pargman, Daniel, and Jacob Palme. "ASCII Imperialism." In *Standards and Their Stories: How Quantifying, Classifying, and Formalizing Practices Shape Everyday Life*, edited by Martha Lampland and Susan Leigh Star, 177–199. Ithaca: Cornell University Press, 2009.

Paris, Britt S., and Morgan Currie. "How the 'Guerrilla Archivists' Saved History—and Are Doing It Again under Trump." *The Conversation*, 21 February 2017. http://theconversation.com/how-the-guerrilla-archivists-saved-history-and-are-doing-it-again-under-trump-72346.

Parkes, Sarah. "Web Content Here Today, Gone Tomorrow." *Canberra Times*, 21 April 2009.

Parman, Molly. "Electronic Archive of the Day US Was Attacked." *CNN*, 6 September 2002.

"Participants." Time & Bits Website. Archived 25 January 1998. Available via the Internet Archive Wayback Machine. https://web.archive.org/web/19980125063112/hhttp://www.gii.getty.edu/timeandbits/part.html.

Pasieczny, Zenon. "A Goldmine in the Sky." *Sydney Morning Herald* (Sydney, AUS). 29 June 1999.

Pearson, David. "The Importance of the Copy Census as a Methodology in Book History." In *Early Printed Books as Material Objects*, edited by Bettina Wagner, 321–328. Berlin: De Gruyter, 2010.

Pew Internet and American Life Project. "One Year Later: September 11 and the Internet." 5 September 2002.

Pew Research. "Online Use." 16 December 1996. https://www.pewresearch.org/politics/1996/12/16/online-use/.

Phillips, Margaret E. "The National Library of Australia: Ensuring Long-Term Access to Online Publications." *Journal of Electronic Publishing* 4, no. 4 (May 1, 1999).

Phillips, Margaret E., and Paul Koerbin. "PANDORA, Australia's Web Archive." *Journal of Internet Cataloging* 7, no. 2 (2004): 19–33.

"Please Help Us Build a Web Archive of the September 11 Attack." webarchivist.org, 20 September 2001. Available via the Internet Archive Wayback Machine. https://web.archive.org/web/20010920003153/http://webarchivist.org/.

Popper, Nathaniel. "Dream of New Kind of Credit Union Is Extinguished by Bureaucracy." *New York Times*, 24 November 2015.

"Preserving the WorldWideWeb." Archived 2 December 1998. Available via the Internet Archive Wayback Machine. https://web.archive.org/web/19981202220930/http:/kulturarw3.kb.se:80/html/preweb.html.

"Proceeds from Sale of AOL Stock Cheque." 17 October 1995. Available at https://archive.org/details/10Kahle000397.

"Proposing a Project or a Donation." Internet Archive, 22 June 2000. Available via Internet Archive Wayback Machine at https://web.archive.org/web/20000622065525/http://archive.org/proposal.html.

ProPublica. "Non-Profit Explorer for the Internet Archive." https://projects.propublica.org/nonprofits/organizations/943242767.

"Public Session." In *Time and Bits: Managing Digital Continuity*, edited by Margaret MacLean and Ben H. Davis, 36–64. Los Angeles: J. Paul Getty Trust, 1998.

Pymm, Bob. "Archives and Web 2.0: The Example of the September 11 Digital Archive." *Archives and Manuscripts* 38, no. 1 (May 2010): 13–26.

Quittner, Joshua. "Invasion of Privacy." *CNN Time*, 25 August 1997. https://www.cnn.com/ALLPOLITICS/1997/08/18/time/pmain.html.

Rainie, Lee, and Bente Kalsnes. "The Commons of the Tragedy: How the Internet Was Used by Millions after the Terror Attacks to Grieve, Console, Share News, and Debate the Country's Response." *Pew Internet and the American Life Project* (10 October 2001): 1–15.

Rankin, Joy Lisi. *A People's History of Computing in the United States*. Cambridge: Harvard University Press, 2018.

Rayward, W. Boyd. *Mundaneum: Archives of Knowledge* (Urbana-Champaign: University of Illinois Graduate School of Library and Information Science, 2010), 3. This book is a translation and adaptation of Raphaèle Cornille, Stéphanie Manfroid, and Manuela Valentino, *Le Mundaneum: les archives de la connaissance* (Brussels: Les impressions Nouvelles, 2008).

Reagan, Brad. "The Digital Ice Age." Popular Mechanics, October 1, 2009. https://www.popularmechanics.com/technology/gadgets/news/4201645.

"Reference Group Meeting." Available via the Internet Archive Wayback Machine. 17 September 1996. https://web.archive.org/web/20021204020028/http://www.kb.se/kw3/Referensgrupp.htm#prot.

"Reference Group Meeting." 11 February 1997. Available via the Internet Archive Wayback Machine. https://web.archive.org/web/20021218220946/http://www.kb.se/kw3/1997-02-11.html.

"Reference Group Meeting." 14 May 1997. Available via the Internet Archive Wayback Machine. https://web.archive.org/web/20021218220614/http://www.kb.se/kw3/1997-05-14.html.

"Reference Group Meeting." 10 September 1997. Available via the Internet Archive Wayback Machine. https://web.archive.org/web/20021218222442/http://www.kb.se/kw3/1997-09-10.html.

"Reference Group Meeting." 3 April 1998. Available via the Internet Archive Wayback Machine. https://web.archive.org/web/20021218222812/http://www.kb.se/kw3/1998-04-03.html.

"Reference Group Meeting." 17 September 1998. Available via the Internet Archive Wayback Machine. https://web.archive.org/web/20021218222620/http://www.kb.se/kw3/1998-09-17.html.

"Reference Group Meeting." 25 November 1999. Available via the Internet Archive Wayback Machine. https://web.archive.org/web/20021218215416/http://www.kb.se/kw3/1999-11-25.html.

"Reference Group Meeting." 8 May 2000. Available via the Internet Archive Wayback Machine. https://web.archive.org/web/20021218215940/http://www.kb.se/kw3/2000-05-08.html.

Rhoads, James B. "Preface." In *The National Archives and Statistical Research*, edited by Meyer H. Fishbein, xi–xii. Athens: Ohio University Press, 1973.

Ridener, John, and Terry Cook. *From Polders to Postmodernism: A Concise History of Archival Theory*. Duluth: Litwin, 2009.

Rivard, Courtney. "Archiving Disaster and National Identity in the Digital Realm: The September 11 Digital Archive and the Hurricane Digital Memory Bank." *Academia.edu* (2011): 1–33. https://www.academia.edu/7455004/Archiving_Disaster_and_National_Identity_in_the_Digital_Realm_The_September_11_Digital_Archive_and_the_Hurricane_Digital_Memory_Bank.

———. "Archiving Disaster and National Identity in the Digital Realm: The September 11 Digital Archive and the Hurricane Digital Memory Bank." In *Identity Technologies: Constructing the Self Online*, edited by Anna Poletti and Julie Rak, 132–143. Madison: University of Wisconsin Press, 2014.

———. "Collecting Disaster: National Identity and the Smithsonian's September 11 Collection." *Australasian Journal of American Studies* 31, no. 2 (December 2012): 87–102.

Robertson, Tara. "Digitization: Just Because You Can, Doesn't Mean You Should." *TaraRobertson.ca*, 20 March 2016. https://tararobertson.ca/2016/oob/.

Robin, Michael. "WAIS—A New Vision for Publishing." *MicroTimes: Northern California's Computer Magazine*, 5 April 1994. https://archive.org/details/05Kahle001743.

Rosenberg, Daniel. "Search." In *Information: A Historical Companion*. Edited by Ann Blair, Paul Duguid, Anja-Silvia Goeing, and Anthony Grafton, 278–279. Princeton: Princeton University Press, 2021.

Rosenzweig, Roy. "Scarcity or Abundance? Preserving the Past in a Digital Era." *American Historical Review* 103, no. 3 (June 2003): 735–762.

Rosenzweig, Roy, Jim Sparrow, and Dan Cohen. "About Us." Echo Project. Archived 16 November 2001. Available via the Internet Archive Wayback Machine. https://web.archive.org/web/20011116204659/http://echo.gmu.edu/about/.

Ross, Seamus. "Historians, Machine-Readable Information, and the Past's Future." In *Electronic Information Resources and Historians: European Perspectives*, edited by Seamus Ross and Edward Higgs, 1–21. St. Katharinen: Max-Planck-Institut für Geschichte in Kommission bei Scripta Mercaturae Verlag, 1993.

———. "Introduction: Networking and Humanities Scholarship." In *Networking in the*

Humanities, edited by Stephanie Kenna and Seamus Ross, xi–xxiv. London: Bowker Saur, 1995.

Rothenberg, Jeff. "Ensuring the Longevity of Digital Documents." *Scientific American* 272, no. 1 (January 1995): 42–47.

Routledge Companion to Global Internet Histories, edited by Gerard Goggin and Mark McLelland. London: Routledge, 2017.

Ruest, Nick, Jimmy Lin, Ian Milligan, and Samantha Fritz. "The Archives Unleashed Project: Technology, Process, and Community to Improve Scholarly Access to Web Archives." In *Proceedings of the ACM/IEEE Joint Conference on Digital Libraries* (2020): 157–166.

Ruest, Nick, and Ian Milligan. "An Open-Source Strategy for Documenting Events: The Case Study of the 42nd Canadian Federal Election on Twitter." *Code4Lib Journal* 32 (2016). https://journal.code4lib.org/articles/11358.

Rumsey, Abby Smith. *When We Are No More: How Digital Memory Is Shaping Our Future*. London: Bloomsbury, 2016.

"safe.millennium.berkeley.edu." University of California, Berkeley, 14 September 2001. Available via the Library of Congress Web Archives. https://webarchive.loc.gov/legacy/20010914180325/http://safe.millennium.berkeley.edu/.

Salmon, Jacqueline L. "Digital 9/11 Project Gets a National Repository." *Washington Post*, 4 September 2003.

Sample, Ian. "Google Boss Warns of 'Forgotten Century' with Email and Photos at Risk." *The Guardian*, February 13, 2015. https://www.theguardian.com/technology/2015/feb/13/google-boss-warns-forgotten-century-email-photos-vint-cerf.

Sanders, Terry, dir., *Into the Future: On the Preservation of Knowledge in the Electronic Age*. 1997. Los Angeles: American Film Foundation, 2018.

Saussy, Haun. "Appraising." In *Information: A Historical Companion*, edited by Ann Blair, Paul Duguid, Anja-Silvia Goeing, and Anthony Grafton, 304–311. Princeton: Princeton University Press, 2021.

"Saving the Histories of September 11, 2001." The September 11 Digital Archive. Archived 17 January 2002. Available via the Internet Archive Wayback Machine. https://web.archive.org/web/20020117115653/http://www.911digitalarchive.org/.

Scasa, Teresa. "A Little Knowledge Is a Dangerous Thing? Information Asymmetries and the Right to Be Forgotten." In *The Right to Be Forgotten: A Canadian and Comparative Perspective*, edited by Ignacio N. Cofone, 26–39. New York: Routledge, 2020.

Schafer, Valérie, Gérôme Truc, Romain Badouard, Lucien Castex, and Francesca Musiani. "Paris and Nice Terrorist Attacks: Exploring Twitter and Web Archives." *Media, War & Conflict* 12, no. 2 (2019): 153–170.

Schürer, Kevin. "Information Technology and the Implications for the Study of History in the Future." In *Electronic Information Resources and Historians: European Perspectives*, edited by Seamus Ross and Edward Higgs, 302–316. St. Katharinen: Max-Planck-Institut für Geschichte in Kommission bei Scripta Mercaturae Verlag, 1993.

———. "The Outlook of the User Community." In *Electronic Information Resources and Historians: European Perspectives*, edited by Seamus Ross and Edward Higgs, 246–248. St. Katharinen: Max-Planck-Institut für Geschichte in Kommission bei Scripta Mercaturae Verlag, 1993.

Schwartz, John. "A Library of Web Pages that Warms the Cockles of the Wired Heart and Beats the Library of Congress for Sheer Volume." *New York Times*, 29 October 2001.

Schweig, Meredith. *Renegade Rhymes: Pop Music, Narrative, and Knowledge in Taiwan*. Chicago: University of Chicago Press, 2022.

Scott, Jason. "Open Source, Open Hostility, Open Doors." Keynote address at Open Source Bridge 2012, 28 June 2012, Portland, OR. Available at https://www.youtube.com/watch?v=tJqZGRIwtxk#t=396s.

———. "Textfiles.com." Undated but accessed 20 May 2022. http://textfiles.com.

Selingo, Jeffrey. "In Attempting to Archive the Entire Internet, a Scientist Develops a New Way to Search It." *Chronicle of Higher Education*, 6 March 1998.

"Senate Amendment to HR 3338." United States Congress, 7 December 2001. https://www.congress.gov/amendment/107th-congress/senate-amendment/2452.

Senk, Sarah. "The Memory Exchange: Public Mourning at the National 9/11 Memorial Museum." *Canadian Review of American Studies* 48, no. 2 (2018): 254–276.

"September 11: Bearing Witness to History." Smithsonian National Museum of American History. Accessed 30 September 2021. https://amhistory.si.edu/september11/.

"September 11 Digital Archive Awarded Saving America's Treasures Grant." *Targeted News Service*, 22 February 2011.

"September 11th Initiative." StoryCorps, accessed 3 October 2020. https://storycorps.org/discover/september-11th/.

"September 11, 2001 Oral History Projects." Columbia University Libraries. Accessed 3 October 2020. https://library.columbia.edu/libraries/ccoh/digital/9-11.html.

"September 11 Web Archive." Minerva: Mapping the Internet Electronic Resources Virtual Archive. 12 April 2005. Available via the Internet Archive Wayback Machine. https://web.archive.org/web/20050412111801/http:/www.loc.gov/minerva/collect/sept11/.

"The 7-Terabyte Man Takes on the Storage Gods." *InternetWeek*, 8 December 1997.

Shepherd, Adam. "Why Facebook Will Plunge Us into a Digital Dark Age." *IT PRO*, March 13, 2019. https://www.itpro.co.uk/mobile/33217/why-facebook-will-plunge-us-into-a-digital-dark-age.

Sinclair, Jenny. "Web Wise." *Sydney Morning Herald*. 11 February 1997.

Skillman, Juanita. "Video Review: Into the Future." *Information Management Journal* (April 1998): 33.

Smith, Wendy. "Still Lost in Cyberspace? Preservation Challenges of Australian Internet Resources." *Australian Library Journal* 54, no. 3 (August 1, 2005): 274–287.

Smithsonian Institute, "National Museum of American History Tracks Presidential Election Process from a Web Perspective." 7 March 1996. Available at https://archive.org/details/04Kahle000463.

Solomon, George. "He Crosses Country to Contest Copyright." *Pittsburgh Post-Gazette*, 10 October 2002.

"Special Release: Post–September 11." UCLA Internet Project Website, 7 February 2002. Available via the Internet Archive Wayback Machine. https://web.archive.org/web/20020802094039/http://ccp.ucla.edu/pages/NewsTopics.asp?Id=32.

Srinivasan, Venkat. "The Internet Archive—Bricks and Mortar Version." *Scientific*

American Blogs, 13 April 2016. https://blogs.scientificamerican.com/guest-blog/the-internet-archive-bricks-and-mortar-version/.
Starr, S. Frederick. *Lost Enlightenment: Central Asia's Golden Age from the Arab Conquest to Tamerlane*. Princeton: Princeton University Press, 2013.
"Statistics." Available via the Internet Archive Wayback Machine. 1 March 2002. https://web.archive.org/web/20021027005825/http://www.kb.se/kw3/Statistik.htm.
Steinberg, Steve G. "Seek and Ye Shall Find (Maybe)." *Wired*, 1 May 1996.
Stephens, Andrew. "Caught in The Web." *The Age* (Melbourne, AUS). 15 May 2010.
Sterling, Bruce. "The Life and Death of Media." Sixth International Symposium on Electronic Art, Montreal, 1995. http://www.servinglibrary.org/journal/1/the-life-and-death-of-media.
Sterling, Bruce, and Richard Kadrey. "Dead Media Project: A Modest Proposal." Accessed October 20, 2020. http://www.deadmedia.org/modest-proposal.html.
Stoddart, Jennifer. "Lost in Translation: Transposing the Right to Be Forgotten from Different Legal Systems." In *The Right to Be Forgotten: A Canadian and Comparative Perspective*, edited by Ignacio N. Cofone, 17–25. New York: Routledge, 2020.
"Stories of September 11." The September 11 Digital Archive. Archived 5 April 2002. Available via the Internet Archive Wayback Machine. https://web.archive.org/web/20020405210414/http://911digitalarchive.org/stories/add.html.
Streitfeld, David. "In a Flood Tide of Digital Data, an Ark Full of Books." *New York Times*, 3 March 2012.
Stross, Randall E. *The Microsoft Way: The Real Story of How the Company Outsmarts Its Competition*. New York: Perseus, 1996.
Suda, Brian. "CERN: Line Mode Browser." optional.is/required, September 25, 2013. http://optional.is/required/2013/09/25/cern-line-mode-browser/.
Sudhir, Pillarisetti. "Preserving the Past: Into the Future with Terry Sanders." *Perspectives on History*, 1 April 1998. https://www.historians.org/publications-and-directories/perspectives-on-history/april-1998/preserving-the-past-into-the-future-with-terry-sanders.
"Summary Video of Key Events on 9/11/2001." Understanding 9/11: A Television News Archive. Updated 24 August 2011. Available via the Internet Archive at https://archive.org/details/911.
Summers, Anthony, and Robbyn Swan. *The Eleventh Day: The Full Story of 9/11*. New York: Ballantine, 2011.
Suzukamo, Leslie Brooks. "E-Mail Archive Holds Initial Reactions of Some to Sept. 11 Attacks." *Saint Paul Pioneer Press*, 9 September 2002.
Swedish Authority for Privacy Protection. "News in English." Updated 27 May 2021. https://www.datainspektionen.se/other-lang/in-english/.
Swierenga, Robert P. "Clio and Computers: A Survey of Computerized Research in History." *Computers and the Humanities* 5, no. 1 (September 1970): 1–21.
Task Force on Archiving of Digital Information. "Preserving Digital Information." 1 May 1996. https://www.clir.org/wp-content/uploads/sites/6/pub63watersgarrett.pdf.
"Thinking Machines in Chapter 11, Big Layoff." *Newsbytes News Network*, 16 August 1994.

Tretikov, Lila. "European Court Decision Punches Holes in Free Knowledge." Wikimedia Foundation, 6 August 2014. https://wikimediafoundation.org/news/2014/08/06/european-court-decision-punches-holes-in-free-knowledge/.

Tsukayama, Hayley. "CERN Reposts the World's First Web Page." *Washington Post*, May 1, 2013. http://www.washingtonpost.com/business/technology/cern-reposts-the-worlds-first-web-page/2013/04/30/d8a70128-b1ac-11e2-bbf2-a6f9e9d79e19_story.html.

Turner, Fred. *From Counterculture to Cyberculture: Stewart Brand, the Whole Earth Network, and the Rise of Digital Utopianism*. Chicago: University of Chicago Press, 2006.

Ubois, Jeff. "It's a Jungle Out There." *Upside*, March 1998.

Vaidhyanathan, Siva. *The Anarchist in the Library: How the Clash Between Freedom and Control Is Hacking the Real World and Crashing the System*. New York: Basic Books, 2004.

Vallis, Mary. "Ordinary Items Become Symbols of Survival." *National Post* (Toronto, ON), 7 September 2002.

"WAIS, Inc. Plan 95 Company Philosophy." 29 June 1994. Available via the Internet Archive Wayback Machine. https://archive.org/details/05Kahle002724/page/n1/mode/2up.

"WAIS Inc. Releases New Network Publishing Software." *Business Wire*, 30 April 1993.

"WAIS: Wide Area Information Server Collection" Available Via the Internet Archive Wayback Machine. https://archive.org/details/wais.

Wallich, Paul. "Preserving the Word." *Scientific American* 278, no. 1 (January 1998): 110.

Walsham, Alexandra. "The Social History of the Archive: Record-Keeping in Early Modern Europe." *Past & Present* 230, suppl. 11 (November 1, 2016): 9–48.

Waters, Donald J. "Choices in Digital Archiving: The American Experience." In *The Impact of Electronic Publishing on the Academic Community*. Edited by Ian Butterworth, 133–141. London: Portland Press, 1998.

———. "Electronic Technologies and Preservation." Paper presented at the Annual Meeting of the Research Libraries Group, 25 June 1992.

———. "How Do We Archive Digital Records? Report of the CPA/RLG Task Force." *DTDA Webpage*. Available via the Internet Archive Wayback Machine. https://web.archive.org/web/19980626160605/http://www.dtda.com/presenta/waters01.htm.

Webster, Peter. "Users, Technologies, Organisations: Towards a Cultural History of World Web Archiving." In *Web 25: Histories from 25 Years of the World Wide Web*, edited by Niels Brügger, 179–190. New York: Peter Lang Publishing, 2017.

Weise, Elizabeth. "Tracking the Web's Beaten Paths." *USA Today*, 17 September 1997.

Wells, H. G. *World Brain*. Cambridge: MIT Press, 2021.

Wendland, Mike. "Overloaded Internet Fails Info-Starved Americans." *Poynter*, 2 September 2002. https://www.poynter.org/archive/2002/overloaded-internet-fails-info-starved-americans/.

"We're Big on Virtual Presence." *Canberra Times*. 10 April 2006.

Wernick, Adam, and Christopher Intagliata. "Scientists Warn We May Be Creating a 'Digital Dark Age.'" *Public Radio International*, January 1, 2018. https://www.pri.org/stories/2018-01-01/scientists-warn-we-may-be-creating-digital-dark-age.

Wikipedia. "List of Websites Founded before 1995." Accessed May 12, 2020. https://en.wikipedia.org/w/index.php?title=List_of_websites_founded_before_1995&oldid=956249448.
Williams, Margot. "Alexa Makes Sense Out of the Chaos that Tangles the Web." *Washington Post*, 28 September 1998.
Wright, Alex. *Cataloguing the World: Paul Otlet and the Birth of the Information Age*. London: Oxford University Press, 2014.
Yale, Elizabeth. "The History of Archives: The State of the Discipline." *Book History* 18 (2015): 332–359.
Yiacoum, Roulla. "In a Pickle." *Sydney Morning Herald*, 13 April 2001.
Young, Jeffrey R. "Scribes of the Digital Era." *Chronicle of Higher Education*, 27 January 2006.
Zweig, Ronald W. "Virtual Records and Real History." *History and Computing* 4, no. 3 (1992): 175.

INDEX

Page numbers in *italic* indicate figures.

Milton Keynes UK
Ingram Content Group UK Ltd.
UKHW041013141124
451058UK00002B/2